Probability, Decisions and Games

A Gentle Introduction using R

Abel Rodríguez
Bruno Mendes

This edition first published 2018
© 2018 John Wiley & Sons, Inc.

The right of Abel Rodríguez and Bruno Mendes to be identified as the authors of this work has been asserted in accordance with law.

Registered Offices
John Wiley & Sons, Inc., 111 River Street, Hoboken, NJ 07030, USA

Editorial Office
111 River Street, Hoboken, NJ 07030, USA

For details of our global editorial offices, customer services, and more information about Wiley products visit us at www.wiley.com.

Wiley also publishes its books in a variety of electronic formats and by print-on-demand. Some content that appears in standard print versions of this book may not be available in other formats.

Library of Congress Cataloging-in-Publication Data:

Names: Rodríguez, Abel, 1975- author. | Mendes, Bruno, 1970- author.
Title: Probability, decisions, and games : a gentle introduction using R / by
    Abel Rodríguez, Bruno Mendes.
Description: Hoboken, NJ : Wiley, 2018. | Includes index. |
Identifiers: LCCN 2017047636 (print) | LCCN 2017059013 (ebook) | ISBN
    9781119302612 (pdf) | ISBN 9781119302629 (epub) | ISBN 9781119302605 (pbk.)
Subjects: LCSH: Game theory–Textbooks. | Game theory–Data processing. |
    Statistical decision–Textbooks. | Statistical decision–Data processing.
    | Probabilities–Textbooks. | Probabilities–Data processing. | R
    (Computer program language)
Classification: LCC QA269 (ebook) | LCC QA269 .R63 2018 (print) | DDC
    519.30285/5133–dc23
LC record available at https://lccn.loc.gov/2017047636

Cover design: Wiley
Cover image: © Jupiterimages/Getty Images

Set in 10/12pt Warnock Pro by SPi Global, Chennai, India

To Sabrina
Abel
To my family
Bruno

Contents

Preface

Why Gambling and Gaming?

Games are a universal part of human experience and are present in almost every culture; the earliest games known (such as *senet* in Egypt or the *Royal Game of Ur* in Iraq) date back to at least 2600 B.C. Games are characterized by a set of rules regulating the behavior of players and by a set of challenges faced by those players, which might involve a monetary or nonmonetary wager. Indeed, the history of gaming is inextricably linked to the history of gambling, and both have played an important role in the development of modern society.

Games have also played a very important role in the development of modern mathematical methods, and they provide a natural framework to introduce simple concepts that have wide applicability in real-life problems. From the point of view of the mathematical tools used for their analysis, games can be broadly divided between random games and strategic games. Random games pit one or more players against "nature" that is, an unintelligent opponent whose acts cannot be predicted with certainty. Roulette is the quintessential example of a random game. On the other hand, strategic games pit two or more intelligent players against each other; the challenge is for one player to outwit their opponents. Strategic games are often subdivided into simultaneous (e.g., rock–paper–scissors) and sequential (e.g., chess, tic-tac-toe) games, depending on the order in which the players take their actions. However, these categories are not mutually exclusive; most modern games involve aspects of both strategic and random games. For example, poker incorporates elements of random games (cards are dealt at random) with those of a sequential strategic game (betting is made in rounds and "bluffing" can win you a game even if your cards are worse than those of your opponent).

One of the key ideas behind the mathematical analysis of games is the rationality assumption, that is, that players are indeed interested in winning the game and that they will take "optimal" (i.e., rational) steps to achieve this. Under these assumptions, we can postulate a theory of how decisions are made, which relies on the maximization of a utility function (often, but certainly not always, related to the amount of money that is made by playing the game). Players attempt to maximize their own utility given the information available to them at any given moment. In the case of random games, this involves making decisions under uncertainty, which naturally leads to the study of probability. In fact, the formal study of probability was born in the seventeenth century from a series of questions posed by an inveterate gambler (Antoine Gambaud, known as the Chevalier de Méré). De Méré, suffered severe financial losses for assessing incorrectly his chances of winning in certain games of dice. Contrary to the ordinary gambler of the time, he pursued the cause of his error with the help of Blaise Pascal, which in turn led to an exchange of letters with Pierre de Fermat and the development of probability theory.

Decision theory also plays an important role in strategic games. In this case, optimality often means evaluating the alternatives available to other players and finding a "best response" to them. This is often taken to mean minimizing losses, but the two concepts are not necessarily identical. Indeed, one important insight gleaned from game theory (the area of mathematics that studies strategic games) is that optimal strategies for zero-sum games (i.e., those games where a player can win only if another loses the same amount) and non zero-sum games can be very different. Also, it is important to highlight that randomness plays a role even in purely strategic games. An excellent example is the game of rock–paper–scissors. In principle, there is nothing inherently random in the rules of this game. However, the optimal strategy for any given player is to select his or her move uniformly at random among the three possible options that give the game its name.

The mathematical concepts underlying the analysis of games and gambles have practical applications in all realms of science. Take for example the game of blackjack. When you play blackjack, you need to sequentially decide whether to *hit* (i.e., get an extra card), *stay* (i.e., stop receiving cards) or, when appropriate, *double down*, *split*, or *surrender*. Optimally playing the game means that these decisions must be taken not only on the basis of the cards you have in your hand but also on the basis of the cards shown by the dealer and all other players. A similar problem arises in the diagnosis and treatment of medical conditions. A doctor has access to a series of diagnostic tests and treatment options; decisions on which one is to be used next needs to be taken sequentially based on the outcomes of previous tests or treatments for this as well as other patients. Poker provides another interesting example. As any experienced player can attest, bluffing is one of the most important parts of the game. The same rules that can be used to decide how to optimally bluff in poker can also be used to design optimal auctions that allow the auctioneer to extract the highest value assigned by the bidders to the object begin auctioned. These strategies are used by companies such as Google and Yahoo to allocate advertising spots.

Using this Book

The goal of this book is to introduce basic concepts of probability, statistics, decision theory, and game theory using games. The material should be suitable for a college-level general education course for undergraduate college students who have taken an algebra or pre-algebra class. In our experience, motivated high-school students who have taken an algebra course should also be capable of handling the material.

The book is organized into 13 chapters, with about half focusing on general concepts that are illustrated using a wide variety of games, and about half focusing specifically on well-known casino games. More specifically, the first two chapters of the book are dedicated to a basic discussion of utility and probability theory in finite, discrete spaces. Then we move to a discussion of five popular casino games: roulette, lotto, craps, blackjack, and poker. Roulette, which is one of the simplest casino games to play and analyze, is used to illustrate the basic concepts in probability such as expectations. Lotto is used to motivate counting rules and the notions of permutations and combinatorial numbers that allow us to compute probabilities in large equiprobable spaces. The games of craps and blackjack are used to illustrate and develop conditional probabilities. Finally, the discussion of poker is helpful to illustrate how many of the ideas from previous chapters fit in together. The last four chapters of the book are dedicated to game theory and strategic games. Since this book is meant to support a general education course, we restrict attention to simultaneous and sequential games of perfect information and avoid games of imperfect information.

The book uses computer simulations to illustrate complex concepts and convince students that the calculations presented in the book are correct. Computer simulations have become a key tool in many areas of scientific inquiry, and we believe that it is important for students to experience how easy access to computing power has changed science over the last 25 years. During the development of the book, we experimented with using spreadsheets but decided that they did not provide enough flexibility. In the end we settled for using R (https://www.r-project.org). R is an interactive environment that allows users to easily implement simple simulations even if they have limited experience with programming. To facilitate its use, we have included an overview and introduction to the R in Appendix A, as well as sidebars in each chapter that introduces features of the language that are relevant for the examples discussed in them. With a little extra work, this book could be used as the basis for a course that introduces students to both probability/statistics *and* programming. Alternatively, the book can also be read while ignoring the R commands and focusing only on the graphs and other output generated by it.

In the past, we have paired the content of this book with screenings of movies from History Channel's *Breaking Vegas* series. We have found the movies *Beat the Wheel, Roulette Attack, Dice Dominator,* and *Professor Blackjack* (each approximately 45 min in length) particularly fitting. These movies are helpful in explaining the rules of the games and providing an entertaining illustration of basic concepts such as the law of large numbers.

November 2017

Abel Rodríguez
Bruno Mendes
Santa Cruz, CA

Acknowledgments

We would like to thank all our colleagues, teaching assistants, and students who thoughtfully helped us to improve our manuscript. In particular, we would like to thank Matthew Heiner and Lelys Bravo for their helpful comments and corrections to earlier drafts of this book. Of course, any inaccuracy is the sole responsibility of the authors.

About the Companion Website

This book is accompanied by a companion website:

www.wiley.com/go/Rodriguez/Probability_Decisions_and_Games

Student Website contains:

- A solutions manual for odd-numbered problems, available for anyone to see.

1

An Introduction to Probability

The study of probability started in the seventeenth century when Antoine Gambaud (who called himself the "Chevalier" de Méré) reached out to the French mathematician Blaise Pascal for an explanation of his gambling loses. De Méré would commonly bet that he could get at least one ace when rolling 4 six-sided dice, and he regularly made money on this bet. When that game started to get old, he started betting on getting at least one double-one in 24 rolls of two dice. Suddenly, he was losing money!

De Méré was dumbfounded. He reasoned that two aces in two rolls are 1/6 as likely as one ace in one roll. To compensate for this lower probability, the two dice should be rolled six times. Finally, to achieve the probability of one ace in four rolls, the number of the rolls should be increased fourfold (to 24). Therefore, you would expect a couple of aces to turn up in 24 double rolls with the same frequency as an ace in four single rolls. As you will see in a minute, although the very first statement is correct, the rest of his argument is not!

1.1 What is Probability?

Let's start by establishing some common language. For our purposes, an *experiment* is any action whose outcome cannot necessarily be predicted with certainty; simple examples include the roll of a die and the card drawn from a well-shuffled deck. The *outcome space* of an experiment is the set of all possible outcomes associated with it; in the case of a die, it is the set $\{1, 2, 3, 4, 5, 6\}$, while for the card drawn from a deck, the outcome space has 52 elements corresponding to all combinations of 13 *numbers* (A, 2, 3, 4, 5, 6, 7, 8, 9, 10, J, Q, K) with four *suits* (hearts, diamonds, clubs, and spades):

$$\{A\heartsuit, 2\heartsuit, 3\heartsuit, 4\heartsuit, 5\heartsuit, 6\heartsuit, 7\heartsuit, 8\heartsuit, 9\heartsuit, 10\heartsuit, J\heartsuit, Q\heartsuit, K\heartsuit,$$
$$A\clubsuit, 2\clubsuit, 3\clubsuit, 4\clubsuit, 5\clubsuit, 6\clubsuit, 7\clubsuit, 8\clubsuit, 9\clubsuit, 10\clubsuit, J\clubsuit, Q\clubsuit, K\clubsuit,$$
$$A\diamondsuit, 2\diamondsuit, 3\diamondsuit, 4\diamondsuit, 5\diamondsuit, 6\diamondsuit, 7\diamondsuit, 8\diamondsuit, 9\diamondsuit, 10\diamondsuit, J\diamondsuit, Q\diamondsuit, K\diamondsuit,$$
$$A\spadesuit, 2\spadesuit, 3\spadesuit, 4\spadesuit, 5\spadesuit, 6\spadesuit, 7\spadesuit, 8\spadesuit, 9\spadesuit, 10\spadesuit, J\spadesuit, Q\spadesuit, K\spadesuit\}$$

Probability, Decisions and Games: A Gentle Introduction using R, First Edition. Abel Rodríguez and Bruno Mendes.
© 2018 John Wiley & Sons, Inc. Published 2018 by John Wiley & Sons, Inc.
Companion website: www.wiley.com/go/Rodriguez/Probability_Decisions_and_Games

A *probability* is a number between 0 and 1 that we attach to each element of the outcome space. Informally, that number simply describes the chance of that event happening. A probability of 1 means that the event will happen for sure, a probability of 0 means that we are talking about an impossible event, and numbers in between represent various degrees of certainty about the occurrence of the event. In the future, we will denote events using capital letters; for example,

A = {It will rain tomorrow},

B = {The number 6 will come up in the next roll of the dice},

while the probability associated with these events is denoted by $P(A)$ and $P(B)$. By definition, the probability of at least one event in the outcome space happening is 1, and therefore the sum of the probabilities associated with each of the outcomes also has to be equal to 1. On the other hand, the probability of an event not happening is simply the *complement* of the probability of the event happening, that is,

$$P(A) = 1 - P(\overline{A})$$

where $\overline{A}$ should be read as "A not happening" or "not A." For example, if A = {It will rain tomorrow}, then $\overline{A}$ = {It will NOT rain tomorrow}.

There are a number of ways in which a probability can be interpreted. Intuitively almost everyone can understand the concept of how likely something is to happen. For instance, everyone will agree on the meaning of statements such as "it is very unlikely to rain tomorrow" or "it is very likely that the LA Lakers will win their next game." Problems arise when we try to be more precise and quantify (i.e., put into numbers) how likely the event is to occur. Mathematicians usually use two different interpretations of probability, which are often called the *frequentist* and *subjective* interpretations.

The frequentist interpretation is used in situations where the experiment in question can be reproduced as many times as desired. Relevant examples for us include rolling a die, drawing cards from a well shuffled deck, or spinning the roulette wheel. In that case, we can think about repeating the experiment a large number of times (call it n) and recording how many of them result in outcome A (call it z_A). The probability of the event A can be defined by thinking about what happens to the ratio $\frac{z_A}{n}$ (sometimes called the empirical frequency) as n grows.

For example, let A = {A flipped coin comes up heads}. We often assign this event a probability of 1/2, that is, we let $P(A) = \frac{1}{2}$. This is often argued on the

---

Sidebar 1.1 Random sampling in R

R provides easy-to-use functions to simulate the results of random experiments. When working with discrete outcome spaces such as those that appear with most casino and tabletop games, the function `sample()` is particularly useful. The first argument of `sample()` is a vector whose entries correspond to the elements of the outcome space, the second is the number of samples that we are interested in drawing, and the third indicates whether sampling will be performed with or without replacement (for now we are only drawing with replacement).

For example, suppose that you want to flip a balanced coin (i.e., a coin that has the same probability of heads and tails) multiple times:

```
> outspc = c("Heads","Tails")        # Outcome space
> z = sample(outspc, 20, replace=TRUE)  # Flip 20 times
> z

 [1] "Tails" "Heads" "Tails" "Tails" "Tails" "Tails" "Heads"
 [8] "Heads" "Tails" "Tails" "Tails" "Heads" "Heads" "Heads"
[15] "Heads" "Tails" "Tails" "Tails" "Heads" "Tails"
```

Similarly, if we want to roll a six-sided die 15 times:

```
> outspc = seq(1,6)
> z = sample(outspc, 15, replace=TRUE)
> z

 [1] 5 2 5 4 2 1 1 2 3 1 6 5 2 6 3
```

---

basis of symmetry: there is no apparent reason why one side of a regular coin would be more likely to come up than the other. Since you can flip a coin as many times as you want, the frequentist interpretation of probability can be used to interpret the value 1/2.

Because flipping the coin by hand is very time-consuming, we instead use a computer to simulate 5000 flips of a coin and plot the cumulative empirical frequency of heads using the following R code (please see Sidebar 1.1 for details on how to simulate random outcomes in R and Figure 1.1 for the output).

```
> n    = 5000
> outc = sample(c("Head","Tail"), n, replace=T)
> z    = cumsum(outc=="Head")/seq(1,n)
> plot(z, xlab="Flips", ylab="Frequency of Heads",type="l")
> abline(h=0.5, col="grey")
```

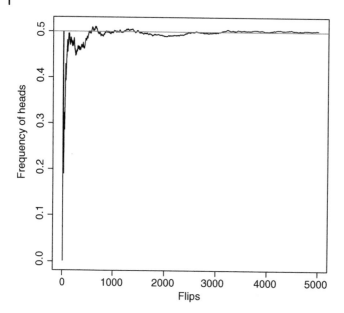

Figure 1.1 Cumulative empirical frequency of heads (black line) in 5000 simulated flips of a fair coin. The gray horizontal line corresponds to the true probability 1/2.

Note that the empirical frequency fluctuates, particularly when you have flipped the coin just a few times. However, as the number of flips (n in our formula) becomes larger and larger, the empirical frequency gets closer and closer to the "true" probability 1/2 and fluctuates less and less around it.

The convergence of the empirical frequency to the true probability of an event is captured by the so-called *law of large numbers*.

Law of Large Numbers for Probabilities

Let z_n represent the number of times that event A happens in a total of n identical repetitions of an experiment, and let $P(A)$ denote the probability of event A. Then $\frac{z_n}{n}$ approaches $P(A)$ as n grows.

This version of the law of large numbers implies that, no matter how rare a non-zero probability event is, if you try enough times, you will eventually observe it. Besides providing a justification for the concept of probability, the law of large numbers also provides a way to compute the probability of complex events by repeating an experiment multiple times and computing the empirical frequency associated with it. In the future, we will do this by using a computer (as we did in our simple coin flipping example before) rather than by physically rolling dice or drawing cards from a deck.

Even though the frequency interpretation of probability we just described is appealing, it cannot be applied to situations where the experiment cannot be repeated. For example, consider the event

$$A = \{\text{It will rain tomorrow}\}.$$

There will be only one tomorrow, so we will only get to observe the "experiment" (whether it rains or not) once. In spite of that, we can still assign a probability to A based on our knowledge of the season, today's weather, and our prior experience of what that implies for the weather tomorrow. In this case, $P(A)$ corresponds to our "degree of belief" on tomorrow's rain. This is a subjective probability, in the sense that two reasonable people might not necessarily agree on the number.

To summarize, although it is easy for us to qualitatively say how likely some event is to happen, it is very challenging if we try to put a number to it. There are a couple of ways in which we can think about this number:

- The frequentist interpretation of probability that is useful when we can repeat and observe an experiment as many times as we want.
- The subjective interpretation of probability, which is useful in almost any probability experiment where we can make a judgment of how likely an event is to happen, even if the experiment cannot be repeated.

1.2 Odds and Probabilities

In casinos and gambling dens, it is very common to express the probability of events in the form of odds (either in favor or against). The *odds in favor* of an event A is simply the ratio of the probability of that event happening divided by the probability of the event not happening, that is,

$$\text{Odds in favor of } A = \frac{P(A)}{1 - P(A)}$$

Similarly, the *odds against A* are simply the reciprocal of the odds in favor, that is,

$$\text{Odds against } A = \frac{1}{\text{Odds in favor of } A} = \frac{1 - P(A)}{P(A)}$$

The odds are typically represented as a ratio of integer numbers. For example, you will often hear that the odds in favor of any given number in American roulette are 1 to 37, or 1:37. Note that you can recover $P(A)$ from the odds in favor of A through the formula,

$$P(A) = \frac{\text{Odds in favor of } A}{1 + \text{Odds in favor of } A}$$

In the context of casino games, the odds we have just discussed are sometimes called the *winning odds* (or the *losing odds*). In that context, you will also hear sometimes about *payoff odds*. This is a bit of a misnomer, as these represent the ratio of payoffs, rather than the ratio of probabilities.

$$\text{Payoff odds in favor of } A = \frac{\text{Payoff to player if } A \text{ happens}}{\text{Payoff to player if } A \text{ does not happen}}$$

For example, the winning odds in favor of any given number in American roulette are 1 to 37, but the payoff odds for the same number are just 1 to 35 (which means that, if you win, every dollar you bet will bring back \$35 in profit). This distinction is important, as many of the odds on display in casinos refer to these payoff odds rather than the winning odds. Keep this in mind!

1.3 Equiprobable Outcome Spaces and De Méré's Problem

In many problems, we can use symmetry arguments to come up with reasonable values for the probability of simple events. For example, consider a very simple experiment consisting of rolling a perfect, six-sided (cubic) die. This type of dice typically has its sides marked with the numbers 1–6. We could ask about the probability that a specific number (say, 3) comes up on top. Since the six sides are the only possible outcomes (we discount the possibility of the die resting on edges or vertexes!) and they are symmetric with respect to each other, there is no reason to think that one is more likely to come up than another. Therefore, it is natural to assign probability 1/6 to each side of the die.

Outcome spaces where all outcomes are assumed to have the same probability (such as the outcome space associated with the roll of a six-sided die) are called equiprobable spaces. In *equiprobable spaces*, the probabilities of different events can be computed using a simple formula:

$$P(A) = \frac{\text{Number of outcomes consistent with } A}{\text{Total number of possible outcomes}}.$$

Note the similarities with the law of large numbers and the frequentist interpretation of probability.

Although the concept of equiprobable spaces is very simple, some care needs to be exercised when applying the formula. Let's go back to Chevalier de Méré's predicament. Recall that De Méré would commonly bet that he could get at least one ace when rolling 4 (fair) six-sided dice, and he would regularly make money on this bet. To make the game more interesting, he started betting on getting at least one double-one in 24 rolls of two dice, after which he started to lose money.

Before analyzing in detail De Méré's bets, let's consider the outcome space associated with rolling two dice. The same symmetry arguments we used in the case of a single die can be used in this case, so it is natural to think of this outcome space as equiprobable. However, there are two ways in which we could construct the outcome space, depending on whether we consider the order of the dice relevant or not (see Table 1.1). The first construction leads to the conclusion that getting a double one has probability $1/21 \approx 0.0476190$, while the second leads to a probability of $1/36 \approx 0.027778$. The question is, which one is the correct one?

In order to gain some intuition, let's run another simulation in R in which two dice are rolled 100,000 times each.

```
> n = 100000
> die1 = sample(seq(1,6), n, replace=T)
> die2 = sample(seq(1,6), n, replace=T)
> sum(die1==1 & die2==1)/n

[1] 0.02765
```

The result of the simulation is very close to $1/36$, which suggests that this is the right answer. A formal argument can be constructed by thinking of the dice as being rolled sequentially rather than simultaneously. Since there

Table 1.1 Two different ways to think about the outcome space associated with rolling two dice.

Order is irrelevant 21 outcomes in total						Order is relevant 36 outcomes in total					
1–1	2–2	3–3	4–4	5–5	6–6	1–1	2–1	3–1	4–1	5–1	6–1
1–2	2–3	3–4	4–5	5–6		1–2	2–2	3–2	4–2	5–2	6–2
1–3	2–4	3–5	4–6			1–3	2–3	3–3	4–3	5–3	6–3
1–4	2–5	3–6				1–4	2–4	3–4	4–4	5–4	6–4
1–5	2–6					1–5	2–5	3–5	4–5	5–5	6–5
1–6						1–6	2–6	3–6	4–6	5–6	6–6

are 6 possible outcomes of the first roll and another 6 possible outcomes for the second one, there is a total of 36 combined outcomes. Since just 1 of these 36 outcomes corresponds to a pair of ones, our formula for the probability of events in equiprobable spaces leads to the probability of 2 ones being 1/36. Underlying this result is a simple principle that we will call the *multiplication principle of counting*,

---

Multiplication Principle for Counting

If events $A, B, C, \ldots$ can each happen in $n_a, n_b, n_c, \ldots$ ways then they can happen together in $n_a \times n_b \times n_c \times \cdots$ ways.

Now, let's go back to De Méré's problem and use the multiplication rule to compute the probability of winning each of his two bets. In this context, it is easier to first compute the probability of losing the bet and, because no ties are possible, then obtain the probability of winning the bet as

$$P(\text{winning}) = 1 - P(\text{losing}).$$

For the first bet, the multiplication rule implies that there are a total of $6 \times 6 \times 6 \times 6 = 6^4 = 1296$ possible outcomes when we roll 4 six-sided dice. If we are patient enough, we can list all the possibilities:

$$1, 1, 1, 1$$
$$1, 1, 2, 2$$
$$1, 1, 1, 3$$
$$1, 1, 1, 4$$
$$1, 1, 1, 5$$
$$1, 1, 1, 6$$
$$1, 1, 2, 1$$
$$\vdots$$

On the other hand, since for each single die there are five outcomes that are not an ace, there are $5^4 = 625$ outcomes for which De Méré losses this bet. Again, we could potentially enumerate these outcomes

$$2, 2, 2, 2$$
$$2, 2, 2, 3$$
$$2, 2, 2, 4$$
$$2, 2, 2, 5$$
$$2, 2, 2, 6$$
$$2, 2, 3, 2$$
$$\vdots$$

The probability that De Méré wins his bet is therefore

$$P(\text{winning first bet}) = 1 - \frac{625}{1296} = \frac{671}{1296} = 0.51775.$$

You can corroborate this result with a simple simulation of 100,000 games:

```
> n = 100000
> die1 = sample(seq(1,6), n, replace=T)
> die2 = sample(seq(1,6), n, replace=T)
> die3 = sample(seq(1,6), n, replace=T)
> die4 = sample(seq(1,6), n, replace=T)
> sum(die1==1 | die2==1 | die3==1 | die4==1)/n

[1] 0.51961
```

For the second bet we can proceed in a similar way. As we discussed before, there are 36 equiprobable outcomes when you roll 2 six-sided dice, 35 of which are unfavorable to the bet. Therefore, there are 36^{24} possible outcomes when two dice are rolled together 24 times, of which 35^{24} are unfavorable to the player, and the probability of winning this bet is equal to

$$P(\text{winning second bet}) = 1 - \frac{35^{24}}{36^{24}} = \frac{36^{24} - 35^{24}}{36^{24}} \approx 0.49140.$$

Again, you can verify the results of the calculation using a simulation:

```
> n = 100000
> outc = seq(1,6)
> numsneakeye = rep(0,n)
> for(i in 1:n){
+     die1 = sample(outc,24,replace=T)
+     die2 = sample(outc,24,replace=T)
+     numsneakeye[i] = sum(die1==1 & die2==1)
+ }
> sum(numsneakeye>=1)/n

[1] 0.49067
```

The fact that the probability of winning is less than 0.5 explains why De Méré was losing money! Note, however, that if he had used 25 rolls instead of 24, then the probability of winning would be $\frac{36^{25}-35^{25}}{36^{25}} \approx 0.50553$, which would have made it a winning bet for De Méré (but not as good as the original one!).

1.4 Probabilities for Compound Events

A *compound event* is an event that is created by aggregating two or more simple events. For example, we might want to know what is the probability that the number selected by the roulette is black or even, or what is the probability that we draw a card from the deck that is both a spade and a number.

As the examples above suggest, we are particularly interested in two types of operations to combine events. On the one hand, the union of two events A and

B (denoted by *A* ∪ *B*) corresponds to the event that happens if either *A* or *B* (or both) happen. On the other hand, the intersection of two events (denoted by *A* ∩ *B*) corresponds to the event that happens only if both *A* and *B* happen simultaneously. The results from these operations can be represented graphically using a Venn diagram (see Figure 1.2) where the simple events *A* and *B* correspond to the rectangles. In Figure 1.2(a), the combination of the areas of both rectangles corresponds to the union of the events. In Figure 1.2(b), the area with the darker highlight corresponds to the intersection of both events. The probability of the intersection of two events is sometimes called the *joint probability* of the two events. In the case when this joint probability is zero (i.e., both events cannot happen simultaneously), we say that the events are *disjoint* or *mutually exclusive*.

In many cases, the probabilities of compound events can be computed directly from the sample space by carefully counting favorable cases. However, in other cases, it is easier to compute them from simpler events. Just as there is

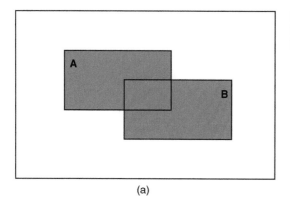

(a)

Figure 1.2 Venn diagram for the (a) union and (b) intersection of two events.

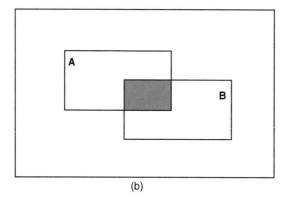

(b)

a rule for probability of two events happening together, there is a second rule for the probability of two alternative events (e.g., the probability of obtaining an even number or a 2 when rolling a die), which is sometimes called the *Addition Rule* of probability:

For any two events,

$$P(A \cup B) = P(A) + P(B) - P(A \cap B)$$

Figure 1.3 presents a graphical representation of two events using Venn diagrams; it provides some hints at why the formula takes this form. If we simply add $P(A)$ and $P(B)$, the darker region (which corresponds to $P(A \cap B)$) is counted twice. Hence, we need to subtract it once in order to get the right result. If two events are *mutually exclusive* (i.e., they cannot occur at the same time, which means that $P(A \cap B) = 0$), this formula reduces to $P(A \cup B) = P(A) + P(B)$.

Similar rules can be constructed to compute the joint probability of two events, $P(A \cap B)$. For the time being, we will only present the simplified *Multiplication Rule* for the probability of independent events. Roughly speaking, this rule is appropriate for when knowing that one of the events occur does not affect the probability that the other will occur.

For any two independent events,

$$P(A \cap B) = P(A)P(B).$$

In Chapter 5, we cover the concept of independent events in more detail and present more general rules to compute the joint probabilities.

Figure 1.3 Venn diagram for the addition rule.

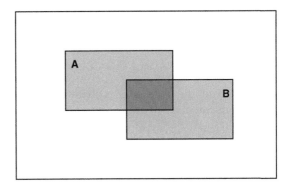

1.5 Exercises

1. A man has 20 shirts and 10 ties. How many different shirt-and-tie combinations can he make?

2. If you have 5 different pants, 12 different shirts, and 3 pairs of shoes, how many days can you go without repeating the same outfit?

3. A fair six-sided die is rolled 1,000,000 times and the number of rolls that turn out to be either a 1 or a 5, $x_{1,5}$ are recorded. From the law of large numbers, what is the approximate value for $x_{1,5}$ that you expect to see?

4. A website asks users to choose eight-letter usernames (only alphabetic characters are allowed, and no distinction is made between lower- and upper-case letters). How many distinct usernames are possible for the website?

5. Provide two examples of experiments for which the probability of the outcomes can only be interpreted from a subjective perspective. For each one of them, justify your choice and provide a value for such probability.

6. In how many ways can 13 students be lined up?

7. Re-write the following probability using the addition rule of probability: P(obtaining a 5 or a 6 when rolling a six-sided die).

8. Re-write the following probability using the addition rule of probability: P(obtaining a total sum of 5 or an even sum when rolling a 2 six-sided dice).

9. Re-write the following probability using the rule of probability for complementary events: P(obtaining at least a 2 when rolling a six-sided die).

10. Re-write the following probability using the rule of probability for complementary events: P(obtaining at most a 5 when rolling a six-sided die).

11. Consider rolling a six-sided die. Which probability rule can be applied to the following probability

   P(obtaining a number higher than 2 or a number smaller than 5).

12. What is the probability of obtaining at least two heads when flipping a coin three times? Which probability rule was used in your reasoning?

13. Explain what is wrong with each of the following arguments.
    (a) First argument:
        - In 1 roll of a six-sided die, I have 1/6 of a chance to get an ace.
        - So in 4 rolls, I have $4 \times \frac{1}{6} = \frac{2}{3}$ of a chance to get at least one ace.
    (b) Second argument:
        - In 1 roll of a pair of six-sided dice, I have 1/36 of a chance to get a double ace.
        - So in 24 rolls, I have $24 \times \frac{1}{36} = \frac{2}{3}$ of a chance to get at least one double ace.

14. What is the probability that, in a group of 30 people, at least two of them have the same birthday. *Hint:* Start by computing the probability that no two people have the same birthday.

15. [R] Write a simulation that allows you to estimate the probability in the previous problem.

16. [R] Modify the code for the second De Méré bet to verify that if 25 rolls are involved instead of 24 then you have a winning bet.

2

Expectations and Fair Values

Let's say that you are offered the following bet: you pay $1, then a coin is flipped. If the coin comes up *tails* you lose your money. On the other hand, if it comes up *heads*, you get back your dollar along with a 50 cents profit. Would you take it?

Your gut feeling probably tells you that this bet is unfair and you should not take it. As a matter of fact, in the long run, it is likely that you will lose more money than you could possibly make (because for every dollar you lose, you will only make a profit of 50 cents, and winning and losing have the same probability). The concept of mathematical expectation allows us to generalize this observation to more complex problems and formally define what a *fair game* is.

2.1 Random Variables

Consider an experiment with numerically valued outcomes $x_1, x_2, \ldots, x_n$. We call the outcome of this type of experiment a *random variable*, and denote it with an uppercase letter such as X. In the case of games and bets, two related types of numerical outcomes arise often. First, we consider the payout of a bet, which we briefly discussed in the previous chapter.

> The *payout* of a bet is the amount of money that is awarded to the player under each possible outcome of a game.

The payout is all about what the player receives after game is played, and it does not account for the amount of money that a player needs to pay to enter it. An alternative outcome that addresses this issue is the profit of a bet:

Probability, Decisions and Games: A Gentle Introduction using R, First Edition. Abel Rodríguez and Bruno Mendes.
© 2018 John Wiley & Sons, Inc. Published 2018 by John Wiley & Sons, Inc.
Companion website: www.wiley.com/go/Rodriguez/Probability_Decisions_and_Games

> The *profit* of a bet is the net change in the player's fortune that results from each of the possible out comes of the game and is defined as the payout minus the cost of entry.
>
> Profit = Payout − Cost of entry

Note that, while all payouts are typically nonnegative, profits can be either positive or negative.

For example, consider the bet that was offered to you in the beginning of this chapter. We can define the random variable

$$X = \{\text{Payout from the bet}\}.$$

As we discussed earlier, this random variable represents how much money a player receives after playing the game. Therefore, the payout has only two possible outcomes $x_1 = 0$ and $x_2 = 1.5$, with associated probabilities $P(X = 0) = 0.5$ and $P(X = 1.5) = 0.5$. Alternatively, we could define the random variable

$$Y = \{\text{Profit from the wager}\},$$

which represents the net gain for a player. Since the price of entry to the game is $1, the random variable Y has possible outcomes $y_1 = -1$ (if the player loses the game) and $y_2 = 0.5$ (when the player wins the game), and associated probabilities $P(Y = -1) = 0.5$ and $P(Y = 0.5) = 0.5$.

2.2 Expected Values

To evaluate a bet, we would like to find a way to summarize the different outcomes and probabilities into a single number. The expectation (or expected value) of a random value allows us to do just that.

> The *expectation* of a random variable X with outcomes $x_1, x_2, \ldots, x_n$ is a weighted average of the outcomes, with the weights given by the probability of each outcome:
>
> $$E(X) = x_1 P(X = x_1) + x_2 P(X = x_2) + \cdots + x_n P(X = x_n).$$

For example, the expected payout of our initial wager is

$$E(X) = \underbrace{(0)}_{x_1} \times \underbrace{(0.5)}_{P(X=x_1)} + \underbrace{(1.5)}_{x_2} \times \underbrace{(0.5)}_{P(X=x_2)} = 0.75.$$

Sidebar 2.1 More on Random Sampling in R.

In Chapter 1, we used the function `sample()` only to simulate outcomes in equiprobable spaces (i.e., spaces where all outcomes have the same probability). However, `sample()` can also be used to sample from nonequiprobable spaces by including a `prob` option. For example, assume that you are playing a game in which you win with probability 2/3, you tie with probability 1/12, and you lose with probability 1/4 (note that 2/3 + 1/12 + 1/4 = 1 as we would expect). To simulate the outcome of repeatedly playing this game 10,000 times, we use

```
> n = 10000
> outsp = c("Win","Tie","Lose")
> x = sample(outsp, n, replace=T, prob=c(2/3,1/12,1/4))
> sum(x=="Win")      # Number of wins in the simulation
```

```
[1] 6589
```

The vector that follows the `prob` option needs to have the same length as the number of outcomes and gives the probabilities associated with each one of them. If the option `prob` is not provided (as in Chapter 1), `sample()` assumes that all probabilities are equal. Hence

```
> x = sample(c("H","T"), n, replace=T, prob=c(1/2,1/2))
```

and

```
> x = sample(c("H","T"), n, replace=T)
```

are equivalent.

On the other hand, the expected profit from that bet is $E(Y) = (-1) \times 0.5 + 0.5 \times 0.5 = -0.25$.

We can think about the expected value as the long-run "average" or "representative" outcome for the experiment. For example, the fact that $E(X) = 0.75$ means that, if you play the game many times, for every dollar you pay, you will get back from the house about 75 cents (or, alternatively, that if you start with $1000, you will probably end up with only about $750 at the end of the day). Similarly, the fact that $E(Y) = -0.25$ means that for every $1000 you bet you expect to lose along $250 (you lose because the expected value is negative). This interpretation is again justified by the law of large numbers:

Law of Large Numbers for Expectations (Law of Averages)
Let $\bar{x}_n = \frac{1}{n}(x_1 + x_2 + \cdots + x_n)$ represent the average outcome of n repetitions of a random variable X with expectation $E(X)$. Then $\bar{x}_n$ approaches $E(X)$ as n grows.

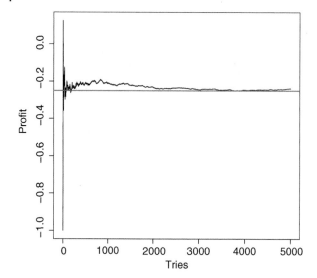

Figure 2.1 Running profits from a wager that costs $1 to join and pays nothing if a coin comes up tails and $1.50 if the coin comes up tails (solid line). The gray horizontal line corresponds to the expected profit.

The following R code can be used to visualize how the running average of the profit associated with our original bet approaches the expected value by simulating the outcome of 5000 such bets and plotting it (see Figure 2.1):

```
> n = 5000
> outcspace = c("Win","Lose")
> res = sample(outcspace, n, replace=T)
> profit = -1*(res=="Lose") + 0.5*(res=="Win")
> runningav = cumsum(profit)/seq(1,n)
> plot(runningav, xlab="Tries", ylab="Profit", type="l")
> abline(h=-0.25, col="grey")
```

The expectation of a random variable has some nifty properties that will be useful in the future. In particular,

> If X and Y are random variables and a, b and c are three constant (non-random numbers), then
>
> $$E(aX + bY + c) = aE(X) + bE(Y) + c.$$

To illustrate this formula, note that for the random variables X and Y we defined in the context of our original bet, we have $Y = X - 1$ (recall our definition of profit and payout minus price of entry). Hence, in this case, we should have $E(Y) = E(X) - 1$, a result that you can easily verify yourself from the facts that $E(Y) = -0.25$ and $E(X) = 0.75$.

2.3 Fair Value of a Bet

We could turn the previous calculation on its head by asking how much money you would be willing to pay to enter a wager. That is, suppose that the bet we proposed in the beginning of this chapter reads instead: you pay me $\$f$, then I flip a coin. If the coin comes up tails, I get to keep your money. On the other hand, if it comes up heads, I give you back the price of bet f along with a 50 cents profit. What is the highest value of f that you would be willing to pay? We call the value of f the *fair value of the bet*.

Since you would like to make money in the long run (or, at least, not lose money), you would probably like to have a nonnegative expected profit, that is, $E(X) \geq 0$, where X is the random variable associated with the profit generated by the bet described earlier. Consequently, the maximum price you would be willing to pay corresponds to the price that makes $E(X) = 0$ (i.e., a price such that you do not make any money in the long term, but at least not lose any either). If the price of the wager is f, then the expected profit of our hypothetical wager is

$$E(X) = -f \times 0.5 + 0.5 \times 0.5 = 0.5 \times (0.5 - f).$$

Note that $E(X) = 0$ if and only if $0.5 - f = 0$, or equivalently, if $f = 0.5$. Hence, to participate in this wager you should be willing to pay any amount equal or lower than the fair value of 50 cents. A game or bet whose price corresponds to its fair value f is called a *fair game* or a *fair bet*.

The concept of fair value of a bet can be used to provide an alternative interpretation of a probability. Consider a bet that pays \$1 if event A happens, and 0 otherwise. The expected value of such a bet is $1 \times P(A) + 0 \times \{1 - P(A)\} = P(A)$, that is, we can think of $P(A)$ as the fair value of a bet that pays \$1 if A happens, and pays nothing otherwise. This interpretation is valid no matter whether the event can be repeated or not. Indeed, this interpretation of probability underlies prediction markets such as PredictIt (https://www.predictit.org) and the Iowa Electronic Market (http://tippie.biz .uiowa.edu/iem/). Although most prediction markets are illegal in the United States (where they are considered a form of online gambling), they do operate in other English-speaking countries such as England and New Zealand.

2.4 Comparing Wagers

The expectation of a random variable can help us compare two bets. For example, consider the following two wagers:

- *Wager 1*: You pay $1 to enter and I roll a die. If it comes up 1, 2, 3, or 4 then I pay you back 50 cents and get to keep 50 cents. If it comes up 5 or 6, then I give you back your dollar and give you 50 cents on top.
- *Wager 2*: You pay $1 to enter and I roll a die. If it comes up 1, 2, 3, 4, or 5 then I return to you only 75 cents and keep 25 cents. If it comes up 6 then I give you back your dollar and give you 75 cents on top.

Let X and Y represent the profits generated by each of the bets above. It is easy to see that, if the dice are fair,

$$E(X) = (-0.5) \times \frac{4}{6} + 0.5 \times \frac{2}{6} \approx -0.166667,$$

$$E(Y) = (-0.25) \times \frac{5}{6} + 0.75 \times \frac{1}{6} \approx -0.083333.$$

These results tell you two things: (1) both bets lose money in the long term because both have negative expected profits; (2) although both are disadvantageous, the second is better than the first because it is the *least negative*.

You can verify the results by simulating 2000 repetitions of each of the two bets using code that is very similar to the one we used in Section 2.2 (see Figure 2.2, as well as Sidebar 2.1 for details on how to simulate outcomes from nonequiprobable experiments in R).

```
> n = 2000
> outsp = seq(1,6)
> die1 = sample(outsp,n,replace=T)
> die2 = sample(outsp,n,replace=T)
> profit1 = 0.5*(die1>4) - 0.5*(die1<=4)
> profit2 = 0.75*(die2>5) - 0.25*(die2<=5)
> runningprf1 = cumsum(profit1)/seq(1,n)
> runningprf2 = cumsum(profit2)/seq(1,n)
> plot(runningprf1, xlab="Tries", ylab="Profit", type="l")
> lines(runningprf2, col="red", lty=2)
```

Note that, although early on the profit from the first bet is slightly better than the profit from the second, once you have been playing both bets for a while the cumulative profits revert to being close to the respective expected values.

Consider now the following pair of bets:

- *Wager 3*: You pay $3 to enter and I roll a die. If it comes up 1, 2, or 3, then I keep your money. If it comes up 4, 5, or 6, then I give you back $6 (your original bet plus a $3 profit).
- *Wager 4*: You pay $3 to enter and I roll a die. If it comes up 1 or 2 then I get to keep your money. If it comes up 3, 4, 5, or 6 then I give you back $4 and a half (your original $3 plus a profit of $1 and a half).

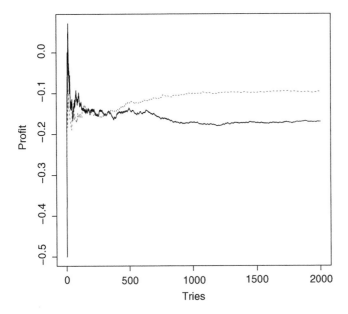

Figure 2.2 Running profits from Wagers 1 (continuous line) and 2 (dashed line).

The expectations associated with these two bets are

$$E(W) = (-3) \times \frac{3}{6} + 3 \times \frac{3}{6} = 0,$$

$$E(Z) = (-3) \times \frac{2}{6} + 1.5 \times \frac{4}{6} = 0.$$

So, both bets are fair, and the expected value does not help us choose among them. However, clearly these bets are not identical. Intuitively, the first one is more "risky", in the sense that the probability of losing our original bet is larger. We can formalize this idea using the notion of *variance* of a random variable:

The *variance* of a random variable X with outcomes $x_1, x_2, \ldots, x_n$ is given by

$$V(X) = E\left\{[X - E(X)]^2\right\} = E(X^2) - \{E(X)\}^2$$
$$= x_1^2 P(X = x_1) + \cdots + x_n^2 P(X = x_n)$$
$$- \{x_1 P(X = x_1) + \cdots + x_n P(X = x_n)\}^2$$

As the formula indicates, the variance measures how far, on average, outcomes are from the expectation. Hence, a larger variance reflects a bet with more extreme outcomes, which often translates into a larger risk of losing money. For instance, for wagers 3 and 4, we have

$$V(W) = \left\{ (-3)^2 \times \frac{3}{6} + 3^2 \times \frac{3}{6} \right\} - \{0\}^2 = 9,$$

$$V(Z) = \left\{ (-3)^2 \times \frac{2}{6} + \left(\frac{3}{2}\right)^2 \times \frac{4}{6} \right\} - \{0\}^2 = 4.5,$$

which agrees with our initial intuition. Figure 2.3 shows the running profit for 2000 simulations of each of the two wagers. As expected, the more variable wager 3 oscillates more wildly and takes longer than the less variable wager 4 to get close to the expected value of 0.

```
> n = 2000
> outsp = seq(1,6)
> die3 = sample(outsp,n,replace=T)
> die4 = sample(outsp,n,replace=T)
> profit3 = 3*(die3>3) - 3*(die3<=3)
> profit4 = 1.5*(die4>2) - 3*(die4<=2)
> runningprf3 = cumsum(profit3)/seq(1,n)
> runningprf4 = cumsum(profit4)/seq(1,n)
> plot(runningprf3, xlab="Tries", ylab="Profit", type="l")
> lines(runningprf4, col="red", lty=2)
> abline(h=0, col="grey")
```

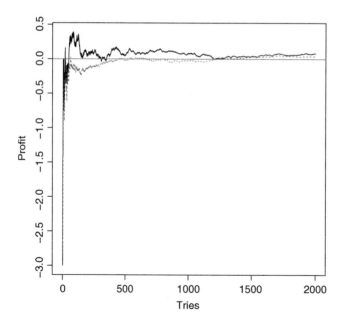

Figure 2.3 Running profits from Wagers 3 (continuous line) and 4 (dashed line).

Just like the expectation, the variance has some interesting properties. First, the variance is always a nonnegative number (a variance of zero corresponds to a nonrandom number). In addition,

---

If X is a random variable and a and b are two constant (non-random numbers), then
$$V(aX + b) = a^2 V(X).$$

A word of caution is appropriate at this point. Note that a larger variance implies not only a higher risk of losing money but also the possibility of making more money in a single round of the game (the maximum profit from wager 3 is actually twice the maximum profit from wager 4). Therefore, if you want to make money as fast as possible (rather than play for as long as you can), you would typically prefer to take an additional risk and go for the bet with the highest variance!

2.5 Utility Functions and Rational Choice Theory

The discussion about the comparison of bets presented in the previous section is an example of the application of rational choice theory. Rational choice theory simply states that individuals make decisions as if attempting to maximize the "happiness" (*utility*) that they derive from their actions. However, before we decide *how to* get what we want, we first need to decide *what* we want. Therefore, the application of the rational choice theory comprises two distinct steps:

1. We need to define a utility function, which is simply a quantification of a person's preferences with respect to certain objects or actions.
2. We need to find the combination of objects/actions that maximizes the (expected) utility.

For example, when we previously compared wagers, our utility function was either the monetary profit generated by the wager (in our first example) or a function of the variance of the wager (when, as the second example, the expected profit from all wagers was the same). However, finding appropriate utility functions for a given situation can be a difficult task. Here are some examples:

1. All games in a casino are biased against the players, that is, all have a negative expected payoff. If the player's utility function were based only on monetary profit, nobody would gamble! Hence, a utility function that justifies people's gambling should include a term that accounts for the nonmonetary rewards associated with gambling.

2. When your dad used to play cards with you when you were five years old, his goal was probably not to win but to entertain you. Again, a utility function based on money probably makes no sense in this case.

3. The value of a given amount of money may depend on how much money you already have. If you are broke, $10,000 probably represents a lot of money, and you would be unwilling to take a bet that would make you lose that much, even if the expected profit were positive. On the other hand, if you are Warren Buffett or Bill Gates, taking such a bet would not be a problem.

In this book, we assume that players are only interested in the economic profit and that the fun they derive from it (the other component of the utility function) is large enough to justify the possibility of losing money when playing. In addition, we will assume that players are risk-averse, so among bets that have the same expected profit, we will prefer those that have lower variances. For this reason, in this book, we will usually look at the expected value of the game first and, if the expected value happens to be the same for two or more choices, we will expect the player pick the one with the lower variance (which, as we discussed before, minimizes the risk).

2.6 Limitations of Rational Choice Theory

Rational choice theory, although useful to formulate models of human behavior, is not always realistic. A good example of how people will easily deviate from the strict rational behavior as defined above is *Ellsberg's Paradox*. Assume that you have an urn that contains 100 blue balls and 200 balls of other colors, some of which are black and some of which are yellow (exactly how many are of each color is unknown). First, you are offered the following two wagers:

- *Wager 1*: You receive $10 if you draw a blue ball and nothing otherwise.
- *Wager 2*: You receive $10 if you draw a black ball and nothing otherwise.

Which of the two wagers would you prefer? After answering this question, you are offered the following two wagers,

- *Wager 3*: You receive $10 if you draw a blue or yellow ball and nothing otherwise.
- *Wager 4*: You receive $10 if you draw a black or yellow ball and nothing otherwise.

No matter how many yellow balls there really are, rational choice theory (based on calculating expected values for each wager) predicts that if you prefer Wager 2 to Wager 1, then you should also prefer Wager 4 to Wager 3, and vice versa. To see this, note that the expected payoff from Wager 1 is $1/3 \times 10 \approx 3.333$ (because there are exactly 100 blue balls in the urn). Consequently, for Wager 2 to be preferable to wager 1, you would need to assume that the urn contains more than 100 black balls. But if you assume that there are at least 100 black balls in the urn, the expected value for Wager 3 would be at most $(100/300 + 99/300) \times 10 \approx 6.663$ (because there are at most 99 yellow balls and exactly 100 blue ball in the urn), while the expected profit for Wager 4 would always be $200/300 \times 10 \approx 6.666$, making Wager 4 always better than Wager 3. The paradox arises from the fact that many people who prefer Wager 1 to Wager 2 actually prefer Wager 4 to Wager 3. This might be because people do not know how to react to the uncertainty of how many black and yellow balls there are and prefer the wagers where there is less (apparent) uncertainty.

Another interesting example is *Allais paradox*. Consider three possible prizes – prize A: $0, prize B: $1,000,000, and prize C: $5,000,000. You are first asked to choose among two lotteries:

- *Lottery 1*: You get prize B ($1,000,000) for sure.
- *Lottery 2*: You get prize A (nothing) with probability 0.01, you get prize B ($1,000,000) with probability 0.89, or you get prize C ($5,000,000) with probability 0.10.

Then you are offered a second set of lotteries

- *Lottery 3*: You get prize A (nothing) with probability 0.89, or you get prize B ($1,000,000) with probability 0.11.
- *Lottery 4*: You get prize A (nothing) with probability 0.90, or you get prize C ($5,000,000) with probability 0.10.

Again, many subjects report that they prefer Lottery 1 to Lottery 2 and Lottery 4 to Lottery 3, although rational choice theory predicts that the persons who choose Lottery 1 should choose Lottery 3 too.

The Allais paradox is even subtler than Elsberg's paradox, because each wager (by itself) has an obvious choice (1 and 4, respectively), but taking the two wagers together, if you choose option 1 in the first wager, you should rationally choose option 3 in the second wager because they are essentially the same option. The way we make sense of this (talk about a paradox!) is by noticing that Lottery 1 can be seen as 89% of the time winning $1 million and the remaining 11% winning $1 million. We look at Lottery 1 in this unusual way because it will be easier to compare it to Lottery 3 (where we win nothing 89% of the time and $1 million 11% of the time). We can change the way we look at Lottery 4 for the same reason (to better compare it to Lottery 2): we win nothing 89% of the time, nothing another 1% of the time, and $5 million 10% of the remaining time. Table 2.1 summaries this alternative description for the lotteries.

You can see that Lotteries 1 and 2 are equivalent 89% of the time (they both give you $1 million) and that Lotteries 3 and 4 are the same also 89% of the time (they give you nothing). Let's look at the table if we cross out the row corresponding to what is supposed to happen 89% of the time.

Table 2.1 Winnings for the different lotteries in Allais paradox.

Lottery 1	Lottery 2	Lottery 3	Lottery 4
Wins $1 million 89% of the time	Wins $1 million 89% of the time	Wins nothing 89% of the time	Wins nothing 89% of the time
Wins $1 million 11% of the time	Wins nothing 1% of the time	Wins $1 million 11% of the time	Wins nothing 1% of the time
	Wins $5 million 10% of the time		Wins $5 million 10% of the time

Table 2.2 Winnings for 11% of the time for the different lotteries in Allais paradox.

Lottery 1	Lottery 2	Lottery 3	Lottery 4
Wins $1 million 11% of the time	Wins nothing 1% of the time	Wins $1 million 11% of the time	Wins nothing 1% of the time
	Wins $5 million 10% of the time		Wins $5 million 10% of the time

In Table 2.2, we see very clearly that Lotteries 1 and 3 are the same choice and that Lotteries 2 and 4 are the same choice too. Hence, the conclusion from this paradox is that by adding winning $1 million 89% of the time in the first wager compared to the second wager, people deviate from the rational choice across wagers even though there's no reason to do so.

The bottom line from these two paradoxes is that, although rational choice is a useful theory that can produce interesting insights, some care needs to be exercised when applying those insights to real life problems, because it seems that people will not necessarily make "rational" choices.

2.7 Exercises

1. Use the definition of rational choice theory to discuss in which sense gambling can be considered "rational" or "irrational."

2. Using the basic principles of the "rational player" described in the text (mainly that a player will always try to maximize its expected value and secondly minimize the variance of the gains), decide which of wagers below would the player choose. In all the wagers, the player is required to pay $1 to enter the wager.

Wager 1: When you flip a coin and heads comes out, you lose your dollar. If tails comes out, you get your dollar back and get an additional $0.25.

Wager 2: When you roll a die and 1 or 2 comes out, you lose your dollar; if a 3 or a 4 comes out, you get your dollar back; and if a 5 or a 5 comes out, you get your dollar back and an additional $0.50.

Wager 3: If you roll a die and 1, 2, or 3 comes out, you lose your dollar. If a 4 comes out, you receive your dollar back, and if a 5 or a 6 comes out, you receive your dollar back and an additional $0.50.

3. The values of random variables are characterized by their random variability. Explain in your own words what aspect of that variability is the expected value trying to capture. What aspect of the random variability is the variance trying to capture?

4. If you are comparing the variance of two different random variables and you find out one is much higher than the other, what does that mean?

5. Does high variability in the profit of a wager mean higher risk or lower risk of losses?

6. The expected profit for a new game with price $1 is −0.0283 cents. If you repeatedly bet $5 for 1000 times, would you expect to win or lose money at the end of the night? How much?

7. Comment on the following statement: "A rational player will always choose a wager with high variability because it allows for higher gains."

8. Consider the three different stocks and their profits. Which one would a rational player choose?

   Stock A: This stock will give you a net profit of $100 with a probability 0.8, a net profit of −$150 with probability 10%, a net profit of $200 with probability 5% or net profit of −$500 with probability 5%.

   Stock B: This stock will give you a net profit of $65 with a probability 0.8, a net profit of −$15 with probability 10%, a net profit of $40 with probability 5% or net profit of −$50 with probability 5%.

   Stock C: This other stock will give you a net profit of $100 with a probability 0.5, a net profit of −$150 with probability 20%, a net profit of $200 with probability 15% or net profit of −$500 with probability 15%.

9. Rank your preferences for the following four lotteries (all cost $1 to enter). Explain your choices:
   - L1: Pays 0 with probability 1/2 and $40,000 with probability 1/2.
   - L2: Pays 0 with probability 1/5 and $25,000 with probability 4/5.

- L3: Pays −$10,000 (so you need to pay $10,000 if you lose!!) with probability 1/2 and $50,000 with probability 1/2.
- L4: Pays $10 with probability 1/3 and $30,000 with probability 2/3.

10. Let's say you finally got some money to buy a decent car. You have two alternatives: alternative U corresponds to buying a 10-old Corolla and alternative N corresponds to buying a brand new Corolla. Each alternative has different types of costs involved (the initial cost of the car and future maintenance costs).

   - For option U, there is a 80% probability that on top of the cost of $10,000 to buy the car we will have a $2000 cost for major work on the car in the future. There is also a probability of 15% for the future costs to be as high as $3000 (for a total cost of $13,000). Finally, the more unlucky are subject to the probability of 5% that future costs be as high as $5000 (for a total cost of $15,000).
   - For option N, there's a pretty high probability (90%) that there are no major costs in maintaining the car in the future and you are subject to just the cost of buying the car ($20,000). There is some chance (say 5%), though, that you might need a new transmission or other major work (say, involving $1000 in costs). There's a smaller probability (3%) of some more serious work being necessary (say something around $2000). And, for the really unlucky, a 2% chance one might need some serious work done (costing something like $3000).

   Which one of the two choices would one rationally recommend and why?

11. An urn contains 30 yellow balls and 70 balls of other colors (which can be either red or blue). Suppose you are offered the following two bets:
   - *Wager 1*: You receive $10 if you draw a yellow ball.
   - *Wager 2*: You receive $10 if you draw a blue ball.

   If you prefer the second bet over the first, which one of the following two wagers would you prefer if you are a rational player?
   - *Wager 3*: You receive $20 if you draw a red ball.
   - *Wager 4*: You receive $15 if you draw a yellow or blue ball.

12. [R] Simulate the profit of both pairs of wagers in the previous exercise and plot the results to see if you made the right decision.

13. A certain health condition has two possible treatments
   - Treatment A, if successful, will extend the lifetime of the patient by 36 months. If it fails, it will neither increase nor reduce the expected lifetime of the patient. Clinical trials show that 20% of patients respond to this treatment.

- Treatment B, if successful, will increase the lifetime of the patient by 14 months, and 65% of the patients respond to it. In addition, 10% of the patients subject to this treatment suffer from an adverse reaction that reduces their expected lifetime by 2 months, and for the rest (25%) the treatment has no effect.

Which one of the two treatments would you recommend, and why? Are there any circumstances under which you would recommend the other treatment? Consider both the point of view of the doctor making the recommendation and that of the patient receiving the treatment.

3

Roulette

Roulette is one of the simpler games available in modern casinos and has captured the popular imagination like no other. In fact, the game has been featured in countless movies such as Humphrey Bogart's 1942 *Casablanca*, Robert Redford's 1993 *Indecent Proposal*, and the 1994 German movie *Run, Lola, Run*.

3.1 Rules and Bets

Roulette is played using a revolving wheel that has been divided into numbered and color-coded pockets. There are 38 pockets in the *American roulette* (popular in the United States), or 37 pockets in the *European roulette* (common in Monte Carlo and other European locations); see Figure 3.1. The *croupier* (as the casino employee in charge of the table is known) spins the wheel and a small ball in opposite directions. The outcome of the game depends on the pocket where the ball falls.

Bets in roulette are placed by moving chips into appropriate locations in the table. Roulette bets are typically divided into *inside* and *outside bets*. Outside bets derive their name from the fact that the boxes where the bets are placed surround the numbered boxes.

The simplest inside bet is called a *straight-up*, which corresponds to a bet made to a specific number. To place this bet, you simply move your chips to the center of the square marked with the corresponding number. The payoff odds from a straight-up bet are 35 to 1, which means that if your number comes up in the wheel, you get your original bet back and get a profit of $35 for each dollar you bet. Other inside bets, such as the *split* or the *street*, are described in Table 3.1.

The simplest of the outside bets are the *color* (or *red/black*) bets, and the *even/odd* bets. As its name indicates, you win a color bet if the ball falls on a

Probability, Decisions and Games: A Gentle Introduction using R, First Edition. Abel Rodríguez and Bruno Mendes.
© 2018 John Wiley & Sons, Inc. Published 2018 by John Wiley & Sons, Inc.
Companion website: www.wiley.com/go/Rodriguez/Probability_Decisions_and_Games

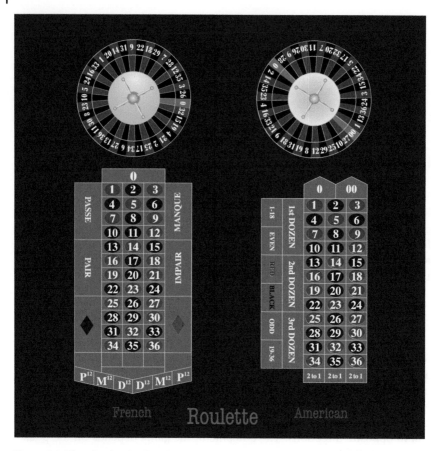

Figure 3.1 The wheel in the French/European (left) and American (right) roulette and respective areas of the roulette table where bets are placed.

pocket that has the same color as the one you picked. Similarly, you win an even bet if the number that comes up in the roulette is a nonzero even number. In both cases, the payoff odds are 1 to 1, so they are often called even bets. However, as we will see below, these even bets are not fair bets because the winning odds are not 1 to 1. A list of outside bets is presented in Table 3.2. This list corresponds to the bets and payoffs most commonly used in the United States; some casinos allow for additional bets, or might slightly change the payouts associated with them.

From a mathematical perspective, the game of roulette is one of the simplest to analyze. For example, in American roulette, there are 38 possible outcomes (the numbers 1–36 plus 0 and 00), which are assumed to be equiprobable. Hence, the probability of any number coming up is 1/38. This means that the

Table 3.1 Inside bets for the American wheel.

Bet name	You are betting on…	Placement of chips	Payout
Straight-up	A single number between 1 and 36	In the middle of number square	35 to 1
Zero	0	In the middle of the 0 square	35 to 1
Double zero	00	In the middle of the 00 square	35 to 1
Split	Two adjoining numbers (horizontally or vertically)	On the edge shared by both numbers	17 to 1
Street	Three numbers on the same horizontal line	Right edge of the line	11 to 1
Square	Four numbers in a square layout (e.g., 19, 20, 22, and 23)	Corner shared by all four numbers	8 to 1
Double street	Two adjoining streets (see Street row)	Rightmost on the line separating the two streets	5 to 1
Basket	One of three possibilities: 0, 1, 2 or 0, 00, 2 or 00, 2, 3	Intersection of the three numbers	11 to 1
Top line	0, 00, 1, 2, 3	Either at the corner of 0 and 1 or the corner of 00 and 3	6 to 1

Table 3.2 Outside bets for the American wheel.

Bet name	You are betting on…	Placement of chips	Payout
Red/black	Which color the roulette will show	Box labeled *Red*	1 to 1
Even/odd	Whether the roulette shows a nonzero even or odd number	Boxes labeled *Even* or *Odd*	1 to 1
1–18	Low 18 numbers	Box labeled 1–18	1 to 1
19–36	High 18 numbers	Box labeled 19–36	1 to 1
Dozen	Either the numbers 1–12 (first dozen), 13–24 (second dozen), or 25–36 (third dozen)	First 12 boxes (first dozen), second 12 boxes (second dozen), third 12 boxes (third dozen).	2 to 1
Column	Either on 1, 4, 7, 10, 13, 16, 19, 22, 25, 28, 31, 34 (left column) or 2, 5, 8, 11, 14, 17, 20, 23, 26, 29, 32, 35 (middle column) or 3, 6, 9, 12, 15, 18, 21, 24, 27, 30, 33, 36 (right column)	Marked box below the corresponding column	2 to 1

expected profit from betting $1 on a straight-up wager is

E(profit on a $1 straight-up bet)

$$= (-1) \times \frac{37}{38} + 35 \times \frac{1}{38} = -\frac{2}{38} \approx -0.0526.$$

Note that the number is negative. Therefore, in the long term, you lose about 5 cents on each dollar you bet. This number is called the *house advantage*, and it ensures that casinos remain a predictably profitable business (remember the law of large numbers for expectation from Chapter 2).

Take now an even bet. There are 18 nonzero even numbers; therefore, the expected profit from this bet is

$$E(\text{profit on a \$1 even bet}) = (-1) \times \frac{20}{38} + 1 \times \frac{18}{38} = -\frac{2}{38} \approx -0.0526.$$

This same calculation applies to odd, red, black, 1–18 and 19–36 bets. On the other hand, for a split bet, we have

$$E(\text{profit on a \$1 split bet}) = (-1) \times \frac{36}{38} + 17 \times \frac{2}{38} = -\frac{2}{38} \approx -0.0526,$$

and for a street bet

$$E(\text{profit on a \$1 street bet}) = (-1) \times \frac{35}{38} + 11 \times \frac{3}{38} = -\frac{2}{38} \approx -0.0526.$$

As a matter of fact, the house advantage for almost every bet in American roulette is the same ($-2/38$). Among the bets discussed in Tables 3.1 and 3.2, the only exception is the top line bet, which is more disadvantageous than the other common bets:

$$E(\text{profit on a \$1 top line bet}) = (-1) \times \frac{33}{38} + 6 \times \frac{5}{38} = -\frac{3}{38} = -0.0789.$$

Indeed, it is more disadvantageous to play a top line bet than to simultaneously bet in each of the numbers included in it using straight-up bets! To see this, consider betting $5 on a top line bet versus betting $1 simultaneously on each of the numbers in the top line bet (0, 00, 1, 2, 3). Due to the properties of expectations (recall Chapter 2), the expected profit from the first wager is

$$\underbrace{5}_{\substack{\text{Dollar} \\ \text{amount} \\ \text{of the} \\ \text{bet}}} \times \underbrace{\left(-\frac{3}{38}\right)}_{\substack{\text{House} \\ \text{advantage on} \\ \text{the top line bet}}} \approx -0.3947,$$

while, the expected profit from the second bet is

$$\underbrace{5}_{\substack{\text{Dollar}\\\text{amount}\\\text{of the}\\\text{bet}}} \times \underbrace{\left(-\frac{2}{38}\right)}_{\substack{\text{House}\\\text{advantage on}\\\text{the straight-up bet}}} \approx -0.2631.$$

This means that with the top line bet you lose on an average 50% more even though you are betting the same amount of money to exactly the same numbers!

Although most bets in roulette are equivalent in terms of their expected value, the risk associated with them differs greatly. Take, for example, the straight-up bet:

$$V(\text{profit on a \$1 straight-up bet})$$
$$= (-1)^2 \times \frac{37}{38} + 35^2 \times \frac{1}{38} - \left(-\frac{2}{38}\right)^2 \approx 33.21,$$

while the variance of a color bet is

$$V(\text{profit on a \$1 color bet})$$
$$= (-1)^2 \times \frac{20}{38} + 1^2 \times \frac{18}{38} - \left(-\frac{2}{38}\right)^2 \approx 0.9972.$$

These calculations highlight that the risk associated with the color bet is much smaller than the risk associated with the straight-up bet. To verify this intuition, let's simulate 10,000 spins of an American roulette and plot the running profit associated with betting \$1 every time on each of the two bets:

```
> n = 10000
> spins = sample (38, n, replace=TRUE)   # 37 and 38 are
>                                         # 0 and 00
> redp = c(1,3,5,7,9,12,14,16,18,19,21,23,
+          25,27,30,32,34,36)   # For the color bet, we are
>                               # betting red, see Figure 3.1
> strp = 16   # For the straight-up bet we are betting to 3
> profit1 = (spins %in% redp) - !(spins %in% redp)
> profit2 = 35*(spins==strp) - (spins!=strp)
> runningav1 = cumsum(profit1)/seq(1,n)
> runningav2 = cumsum(profit2)/seq(1,n)
> plot(runningav1, xlab="Spin", ylab="Profit", type="l")
> lines(runningav2, col="red", lty=2)
> abline(h=-2/38, col="grey")
```

Figure 3.2, which shows the results from these simulations, is consistent with the discussion we had before: although the average profits from both bets eventually tends to converge toward the expected value of $-2/38$, the straight-up bet (which has the highest variance) has much more volatile returns.

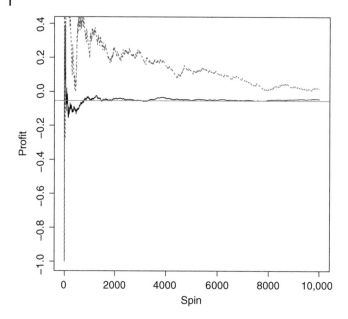

Figure 3.2 Running profits from a color (solid line) and straight-up (dashed line) bet.

What should you do next time you visit a casino? It depends on what your goal is, and how much money you have in your bankroll. If you want to play for as long as possible before exhausting your money, you should only play color bets (or bets with similar payouts such as the even/odd bet or the 1–18 or 19–38 bets). On the other hand, if you want to maximize the amount of money you can potentially make from a single bet, then you should play only straight-up bets (just like the main character in *Run Lola, Run*). However, in that case, you are also maximizing the probability that you will go bankrupt very quickly. There are no free lunches…

European roulette can be analyzed in a similar way. Payout odds are the same as in American roulette, but now the probability of a single number is $\frac{1}{37}$. Hence, the expected value of a straight-up bet is in this case

$$E(\text{profit on a \$1 bet on straight-up})$$
$$= (-1) \times \frac{36}{37} + 35 \times \frac{1}{37} = -\frac{1}{37} \approx -0.027027.$$

This calculation shows that the house advantage for straight-up bets in American roulette is almost twice as large as for European roulette. If you have a choice, you should always prefer roulette in Europe rather than in the United States.

3.2   Combining Bets

Sometimes players like to place multiple bets simultaneously on the same spin of the roulette wheel. As an example, consider placing $2 on a bet of red and $1 on the second dozen. The payout of this simultaneous bet is going to be different depending on the number that comes up. If a red number among the second dozen comes up (such as 16), you win both bets and you get back the $3 you originally bet plus a profit of $4 more (recall that the payoff odds for reds is 1 to 1, which means that you profit $1 for each dollar you bet, while the payoff of a dozen bet is 1 to 2, meaning that you profit $2 for each dollar you bet). On the other hand, if an odd number in the second dozen comes up you lose the even bet, but you win the dozen bet. Table 3.3 describes the possible outcomes along with their probabilities, payouts, and profits (remember that the entry cost of the bet is always $3).

From Table 3.3, it is easy to see that the expected profit of this combined bet is

$$E(\text{profit}) = 4 \times \frac{6}{38} + 0 \times \frac{6}{38} + 1 \times \frac{12}{38} + (-3) \times \frac{14}{38} = -\frac{6}{38}.$$

This is the same expected profit that you would get from betting $3 on pretty much any simple bet! This result suggests that you cannot decrease the house advantage by combining bets. On the other hand, the variance of this bet is

$$V(\text{profit}) = (4)^2 \times \frac{6}{38} + (0)^2 \times \frac{6}{38} + (1)^2 \times \frac{12}{38} + (-3)^2 \times \frac{14}{38} - \left(\frac{6}{38}\right)^2$$
$$= 6.13296.$$

Note that this value is smaller than the variance of the dozen bet, but larger than the variance of the color bet. Therefore, even though it will not increase your probability of winning, mixing bets allows you to tailor the risk that you assume.

Table 3.3 Outcomes of a combined bet of $2 on red and $1 on the second dozen.

Outcome	Prob	Payout ($)	Profit ($)
Red in second dozen (14, 16, 18, 19, 21, 23)	6/38	$2 + 2 + 1 + 2 = 7$	$7 - 3 = 4$
Black in second dozen (13, 15, 17, 20, 22, 24)	6/38	$1 + 2 = 3$	$3 - 3 = 0$
Red in first or third dozen (1, 3, 4, 5, 9, 12, 25, 27, 30, 32, 34, 36)	12/38	$2 + 2 = 4$	$4 - 3 = 1$
All other numbers (0, 00, 2, 6, 7, 8, 10, 11, 26, 28, 29, 31, 33, 35)	14/38	0	$0 - 3 = -3$

3.3 Biased Wheels

So far, our analysis of the game of roulette has assumed that all numbers have the same probability. However, in reality, the wheel is a mechanical device subject to wear and tear so, in time, even the best roulette tends to slightly favor some numbers (in other words, it becomes biased). For example, this might happen because of warped axels, because of chipped/battered pockets, or simply because the wheel is not level. Alternatively, the bias might not be due to the wheel itself, but to the way the croupier spins the wheel or the ball. Whatever be the reason, a biased roulette hurts the casino only if players are aware of it, in which case they can exploit the bias to reduce (or eliminate) the house advantage.

Let's consider the analysis of a biased wheel. For the sake of concreteness, we will work with a wheel that has a slightly positive bias toward three numbers and a negative bias toward the others. A player who knows which three numbers have a positive bias can exploit it by making simultaneous straight-up bets to these numbers. For example, consider a biased wheel in which the numbers 2, 4, and 21 each have a probability of 0.028 (which is slightly larger than the typical $1/38 \approx 0.0263$ associated with an unbiased wheel), while the other 35 numbers have all the same probability of 0.02617143 of coming up (which is necessarily slightly lower than $1/38 \approx 0.0263$). If the player only makes straight-up bets of $1 to each of the three numbers favored by the wheel, the profit of the bet is

$$\underbrace{0.028}_{\substack{\text{Probability of} \\ \text{obtaining a 4}}} \times (\underbrace{36}_{\text{Payout of 4}} - \underbrace{3}_{\text{Cost}}) + \underbrace{0.028}_{\substack{\text{Probability of} \\ \text{obtaining a 21}}} \times (\underbrace{36}_{\text{Payout of 21}} - \underbrace{3}_{\text{Cost}})$$

$$+ \underbrace{0.028}_{\substack{\text{Probability of} \\ \text{obtaining a 2}}} \times (\underbrace{36}_{\text{Payout of 2}} - \underbrace{3}_{\text{Cost}}) + \underbrace{(1 - 3 \times 0.028)}_{\substack{\text{Probability} \\ \text{of other} \\ \text{outcomes}}} \underbrace{(-3)}_{\text{Cost}},$$

which reduces to

$$0.084 \times 33 + 0.916 \times (-3) = 0.024.$$

Since the expected value is positive, the player will actually make money in the long run by betting on the numbers favored by this biased roulette! The fact a player can potentially exploit any bias in the wheel means that casinos are unlikely to bias their wheels on purpose.

Now, let's turn the previous calculation around to answer the following question: how large does the combined bias of the three numbers needs to be in order to eliminate the house advantage (and make the game of roulette fair) if the player is able to discover it? Let x be the combined bias and assume that we bet $1 on each number. That means that the probability of winning with the simultaneous straight up bet is $3/38 + x$ (with a profit of $ 36 - 3 = 33$), and the probability of losing is $35/38 - x$ (with a expected profit of $\$ -3$). The expected profit is therefore

$$33 \times \left(\frac{3}{38} + x\right) + (-3) \times \left(\frac{35}{38} - x\right) = \frac{99 - 105}{38} + 36x = -\frac{6}{38} + 36x,$$

and to make the game fair we need x to satisfy $-6/38 + 36x = 0$, or $x = 1/228$. Therefore, it is enough to change the probability of three numbers from $1/38 \approx 0.026315$ to $1/38 + 1/684 \approx 0.02777$ to make the house edge disappear.

We can use R to simulate a biased wheel in which the numbers 2, 4, and 21 are favored:

```
> n = 5000
> outspc = seq(1,38)
> pbw = rep(0,38)
> fav = c(2,4,21)      # Favored pockets
> pbw[fav]   = 0.028     # Probability of favored pockets
> pbw[-fav]  = (1-sum(pbw[fav]))/35   # Probability of other
> spins = sample(38, n, replace=TRUE, prob=pbw)
> barplot(table(spins))
```

Figure 3.3 presents a bar plot of the observed empirical frequencies observed in 5000 spins of a biased wheel coming out of the simulation above. Note that even with 5000 spins it is not easy to identify which pockets (if any) are favored. Indeed, from this simulation, it would appear that the wheel is biased toward the number 28!

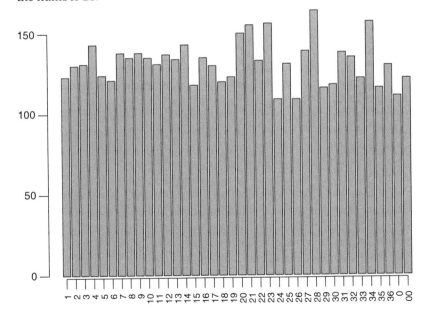

Figure 3.3 Empirical frequency of each pocket in 5000 spins of a biased wheel.

The previous simulation highlights that, even though exploiting a biased wheel is one of the very few ways in which a player can reduce the house edge, detecting a biased wheel with a good degree of certainty is not easy and might require that we observe the wheel for a really long time (particularly if the bias is small). The exact number of spins to be recorded depends both on the size of the bias and on how certain you want to be about the existence of the bias; a rough approximation of the necessary number can be obtained using *Chebyshev's theorem*.

Let z_A/n be the observed frequency of event A after n identical repetitions of an experiment, and let $P(A)$ be the probability associated with the event A. Then, for any desired precision ε,

$$P\left(\left|\frac{z_A}{n} - P(A)\right| > \varepsilon\right) \leq \frac{P(A)\{1 - P(A)\}}{n\varepsilon^2}.$$

Chebyshev's theorem links the number of repetitions of an experiment (in our case, the number of spins) with the error committed when approximating $P(A)$ by the empirical frequency z_A/n. We already knew (from the law of large numbers) that this error becomes smaller and smaller as n grows, but we had no idea about exactly how fast it decreased. Chebyshev's theorem fills that gap, and it can help us determine how many spins are needed to figure out if a wheel is biased or not.

Understanding Chebyshev's theorem can be hard because there are so many frequencies and probabilities involved in the definition. To gain some intuition, consider running a large number of simulations of an unbiased wheel, each consisting of 10,000 spins. Figure 3.4 shows curves for the cumulative empirical frequency of any given pocket for each of 100 such simulations. Because these are random experiments, each curve is slightly different from the others. In spite of this, some patterns are clear. For example, you can see that the graph looks a little bit like a horizontal funnel, wider on the left size and narrower on the right.

We can relate the features of the graph with the different terms that appear in Chebyshev's theorem. For example, you can think about the width of the funnel as the error that is committed when we approximate $P(A)$ by the empirical frequency z_A/n. Hence, the width of the funnel is, roughly speaking, equivalent to ϵ. As expected, smaller values of ϵ (more precision in the estimation) requires larger values of n, and vice versa. Furthermore, it should be clear that, for any n, the width of the funnel is itself random. Indeed, if we do a second set of 100 simulations, the width of the funnel will be slightly different. Hence, the best we can do is to select an n that will give us the desired width *with high probability* (but we can never be absolutely sure that the error is not bigger than we want). The desired high probability is something we need to decide ourselves

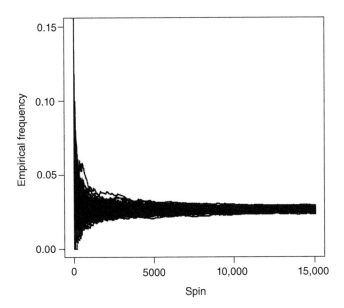

Figure 3.4 Cumulative empirical frequency for a single pocket in an unbiased wheel.

(usually 0.95, 0.99, or 0.999 are used). Note, however, that the larger the desired probability, the larger the value of n needs to be.

To illustrate how Chebyshev's theorem works, consider spinning a wheel 100,000 times to determine whether the probability of 00 is $1/38$. If the wheel is unbiased, how large is the probability that the difference between the estimate we get from this experiment and the true probability is greater than 0.001 (which is the maximum error that I am willing to admit)? A direct application of Chebyshev's theorem yields:

$$P\left(\left|\frac{z_A}{n} - \frac{1}{38}\right| > 0.001\right) \leq \frac{\frac{1}{38} \times \frac{37}{38}}{100{,}000 \times (0.001)^2} \approx 0.2562.$$

This probability is relatively high, so we actually need a much larger number of spins to be accurate enough. How many more? Let's say that I do not want the probability of a 0.001 error to be more than 5%. Then, again from Chebyshev's theorem,

$$0.05 = \frac{\frac{1}{38} \times \frac{37}{38}}{n \times (0.001)^2},$$

which implies that

$$n = \frac{\frac{1}{38} \times \frac{37}{38}}{0.05 \times (0.001)^2} \approx 512{,}466.$$

3.4 Exercises

1. What are even bets in roulette? Are they really even?

2. Albert Einstein once said that no one could possibly win at roulette "unless he steals money from the table while the croupier isn't looking." Explain this statement in the context of the law of large numbers.

3. What is the expected value and variance for simultaneous $5 street-bet on 22–23–24 and a $1 odd-bet in American roulette?

4. What is the expected value and variance for a simultaneous $2 square-bet on 1–2–4–5 and a $3 bet on the first column in American roulette?

5. What is the house advantage in European roulette for the split, color, and dozen bets?

6. In American roulette, almost all bets have the same expected value. However, do they all have the same variance? If you think that they do not, you should show a counterexample. If you want to play for as long as possible, what bet would you prefer?

7. What would be the house advantage in roulette if there were three "losing" numbers in the wheel (call them 0, 00, and 000)?

8. In American-style roulette, which one of the following bets has a higher winning probability? Which one has a higher expected payoff? Which one has the highest variance? Which one would you prefer, and why?
   - Bet $18 on red.
   - Bet $2 on a split.

9. In American-style roulette, which one of the following bets has a higher winning probability? Which one has a higher expected payoff? Which one has the highest variance? Which one would you prefer, and why?
   - Bet $10 on a double-street.
   - Bet $2 on a dozen (assume that the dozen does not contain any number from the double-street).

10. In the *first and third column strategy* in roulette, one bets two pieces in the first column, two pieces in the third column and two pieces in black. What is the expected value of this system in American roulette?

11. Consider a roulette wheel that is positively biased toward two numbers, 9 and 34. How large does the bias need to be in order for bets on this roulette to be fair?

12. Why are casinos unlikely to bias their roulette wheels on purpose?

13. The payoffs in roulette are selected assuming that all numbers have the same probability (in European roulette, i.e., $1/37 \approx 0.027$). Assume that, after collecting many spins from one given European roulette, you find that three numbers, 25, 17, and 34 have a slightly higher probability of coming out (say 0.04), while the other 34 numbers have about the same probability ($0.88/34 \approx 0.02588$). Assume that you pick a strategy where you make $1 straight up bets to each one of the high probability numbers. What would be the expected payoff from this bet?

14. Making the same assumptions as the previous question, but now assume that the three biased numbers have a probability of 0.05, and the remaining numbers have a probability of $0.85/34 \approx 0.025$. Assume that you pick a strategy where you make $1 straight up bets to each one of the high probability numbers. What would be the expected payoff from this bet?

15. How many spins of the wheel should you observe in order to be 99% sure that there is a bias of 0.0029238 in the usual probability for a single number (0.02163157) in an American roulette wheel (this is a bias that not only would cancel the house advantage but also would actually turn it into your advantage).

16. [R] Simulate and visualize the expected value and spread of the even bet in roulette.

17. [R] Simulate a biased roulette, where three numbers of your choice have a probability of occurring 0.3. Produce a histogram of the simulations.

4

Lotto and Combinatorial Numbers

Lotteries are a very common and popular type of gamble. The bet is typically targeted to a number (or set of numbers), which are randomly selected during a special event held one or more times a week. In most cases, the randomization mechanism used in lotteries is such that any number has the same probability of occurring so we are dealing with an equiprobable outcome space. In the United States, state governments typically administer lotteries, and the profits they generate are used to provide supplemental funding to public programs such as public schools and colleges; for example, see http://www.calottery .com/default.htm.

4.1 Rules and Bets

Lotto is one variant of lotteries that is extremely popular. In its simplest version, the player pays a fixed price for a ticket (often $1) and gets to select a few numbers (often 5 or 6) from among a longer list of them (anywhere between 42 and 90 numbers). You win different prizes depending on how many numbers in your list match the random draw (the more matches, the larger the prize). To reduce the risk to the organization running the lottery, the prizes are often set as a fixed percentage of total revenue (often a 50–50% split is used, making this type of game an extremely bad proposition, at least when compared with most casino games). A casino version of lotto, often called *keno*, is also widely popular but is even more difficult to win than the version offered by most state lotteries.

4.1.1 The Colorado Lotto

To be more specific, consider the 6-of-42 lotto offered by the Colorado Lottery (http://www.coloradolottery.com/GAMES/LOTTO/), which is drawn twice a week on Wednesday and Saturday evenings. This game costs $1 to play, distributes 50% of the revenue and pays out if you match 3, 4, 5, or 6 of the

Probability, Decisions and Games: A Gentle Introduction using R, First Edition. Abel Rodríguez and Bruno Mendes.
© 2018 John Wiley & Sons, Inc. Published 2018 by John Wiley & Sons, Inc.
Companion website: www.wiley.com/go/Rodriguez_Probability_Decisions_and_Games

drawn numbers. We would like to compute the probability of winning for each of these four prizes.

Since we are dealing with an equiprobable space, we use the formula,

$$P(\text{winning}) = \frac{\text{Number of groups of 6 numbers that make you win}}{\text{Number of possible groups of 6 numbers (out of 42)}}.$$

Let's start with the probability of winning the main prize (i.e., of picking the 6 right numbers). The numerator in this case is very easy; since the order in which the numbers come up does not matter, there is just one combination of numbers that allows you to win.

To figure out what the denominator is, let's start by using the multiplication rule. We need to choose six numbers without replacement (i.e., the numbers cannot be repeated). Therefore, we have 42 options for the first number, 41 for the second, 40 the third, and so on. This means that, as a first approximation, we have:

$$\begin{array}{l}\text{Number of possible sets}\\ \text{of 6 numbers (out of 42)}\end{array} = 42 \times 41 \times 40 \times 39 \times 38 \times 37 = 3{,}776{,}965{,}920.$$

This number is known as the number of *permutations* of 42 objects taken 6 at a time, and is denoted as $_{42}P_6$. This number can also be computed as

$$_{42}P_6 = \frac{42 \times 41 \times 40 \times 39 \times 38 \times 37 \times 36 \times 35 \times \cdots \times 1}{36 \times 35 \times \cdots \times 1} = \frac{42!}{(42-6)!},$$

where $n! = n \times (n-1) \times (n-2) \times \cdots \times 3 \times 2 \times 1$ is read *n factorial* (with the convention $0! = 1$).

When computing the number of permutations, we implicitly assume that the order in which the numbers appear is important. That is, we assume that the sequences $\{18, 4, 16, 32, 44, 37\}$ and $\{16, 32, 44, 4, 18, 37\}$ are different. However, for the purpose of the game of lotto, these two sequences are really the same. To adjust for this, we need to figure out how many different orderings we have for 6 numbers. Again, we can use the multiplication rule: we need to fill up 6 spots with 6 numbers, so there are 6 options for the first number, 5 spots for the second, and so on. Hence, the total number of ordering of 6 numbers is $_6P_6 = 6 \times 5 \times 4 \times 3 \times 2 \times 1 = 720$.

Since our previous calculation was counting each different combination of 6 numbers 720 times, we just need to divide $_{42}P_6 = 3{,}776{,}965{,}920$ (the number of subsets in which the permutation is important) by the total numbers of ways we can order 6 digits ($_6P_6 = 720$),

$$\begin{array}{l}\text{Number of possible sets of 6}\\ \text{numbers (out of 42) that ignore}\\ \text{the ordering of the elements}\end{array} = \frac{42 \times 41 \times 40 \times 39 \times 38 \times 37}{6 \times 5 \times 4 \times 3 \times 2 \times 1} = 5{,}245{,}786,$$

and the probability of matching exactly the 6 winning numbers

$$P(\text{winning first prize}) = \frac{1}{5{,}245{,}786} \approx 0.00000019063.$$

Sidebar 4.1 Counting Permutations in R

The R function `factorial()` gives you an easy to way to compute (surprise!) factorials.

```
> factorial(6)   # 6! = 6*5*4*3*2*1 = 720

[1] 720
```

Although R does not have a special function to compute the number of permutations of n objects taken in groups of m, you can use `factorial()` and the formulas in this chapter to achieve the same goal. For example, to compute $_{42}P_6$ you can do

```
> factorial(42)/factorial(36)   # n!/(n-m)!

[1] 3776965920
```

You must be careful, however, as this might fail when n and m are large. A trick to get around this is to use the `lfactorial()` function (which directly computes the logarithm of the factorial) and use rules for logarithms and exponents. For example, if you need to compute $_{400}P_2 = 400 \times 399 = 159,600$,

```
> factorial(400)/factorial(398)   # Fails

Warning in factorial(400): value out of range in 'gammafn'
Warning in factorial(398): value out of range in 'gammafn'

[1] NaN

> exp(lfactorial(400) - lfactorial(398))   # Works

[1] 159600
```

Note that the whole calculation for the number of possible groups of 6 numbers out of 42 can be written as

$$\frac{42 \times 41 \times 40 \times 39 \times 38 \times 37}{720}$$

$$= \frac{42 \times 41 \times \cdots \times 3 \times 2 \times 1}{(36 \times 35 \times \cdots \times 3 \times 2 \times 1)(6 \times 5 \times \cdots \times 2 \times 1)} = \frac{42!}{36! \times 6!}$$

This quantity is known as the number of *combinations* of 42 elements taken 6 at a time, and is denoted as

$$\binom{42}{6} \quad \text{or} \quad _{42}C_6,$$

which is read *42 choose 6*.

Sidebar 4.2 Counting Combinations in R

R does provide a specialized function `choose()` for computing the number combinations of n objects taken in groups of m. For example, $\binom{42}{6}$ can be computed as

```
> choose(42,6)

[1] 5245786
```

   Although we know how to count the number of combinations (and permutations!), it might sometimes be useful to go a little further and fully enumerate them. The package `prob` provides the function `urnsamples`, which does exactly that. For example, to enumerate all combinations of 5 elements taken in groups of 3 (there are 10 of them):

```
> library(prob)     # Remember to load the library first!
> elem = seq(1,5)
> urnsamples(elem, size=3, replace=FALSE, ordered=FALSE)

   X1 X2 X3
1   1  2  3
2   1  2  4
3   1  2  5
4   1  3  4
5   1  3  5
6   1  4  5
7   2  3  4
8   2  3  5
9   2  4  5
10  3  4  5
```

   The first argument of `urnsample` defines the set of elements from the groups will be selected (in this case, the numbers between 1 and 5). The second element is the size of the subgroup (3 in this case), and the last two control whether elements can be reused (in which case `replace` should be `TRUE`) and whether order matters (in which case `ordered` should be `TRUE` too).

   The results discussed earlier can be extended to situations in which m objects need to be chosen from among n of them.

The number of ways in which m *ordered* objects can be selected from a total of n options is given by the permutation number

$$_nP_m = \frac{n!}{(n-m)!} = \frac{n \times (n-1) \times (n-2) \times \cdots \times 2 \times 1}{(n-m) \times \cdots \times 2 \times 1}.$$

In the special case where we are interested in the number of ways n objects are ordered this reduces to

$$_nP_n = n! = n \times (n-1) \times (n-2) \times \cdots \times 2 \times 1.$$

The number of ways in in which m *unordered* objects can be selected from a total of n options is given by the combinatorial number (sometimes called the binomial coefficient)

$$_nC_m = \binom{n}{m} = \frac{n!}{(n-m)! \times m!}$$
$$= \frac{n \times (n-1) \times (n-2) \times \cdots \times 2 \times 1}{\{(n-m) \times \cdots \times 2 \times 1\} \times \{m \times \cdots \times 2 \times 1\}}.$$

To convince yourself that the formula for the combinatorial number is correct, consider a simple example in which we want to enumerate all the possible options. In particular, let's compute $\binom{6}{3}$, the number of ways in which 3 numbers can be selected out of 6 without repetition. Our formula above says that

$$\binom{6}{3} = \frac{6!}{3! \times 3!} = \frac{6 \times 5 \times 4 \times 3 \times 2 \times 1}{(3 \times 2 \times 1) \times (3 \times 2 \times 1)} = 20.$$

This can be verified by explicitly enumerating all possible options (see Table 4.1). Sidebars 4.1 and 4.2 discuss how to use R to enumerate and count permutations and combinations.

Let's proceed now to calculate the probability of winning the second prize, that is, matching exactly 5 numbers out of 6. The denominator is the same as before, so we do not need to repeat the calculation. For the numerator, we need to pick 5 numbers out of the 6 that came up in the drawing, while the sixth number needs to come up from among the 36 that are not winning numbers. So the numerator is

$$\binom{6}{5} \times 36 = \frac{36 \times 6!}{(6-5)! \times 5!} = \frac{36 \times 6 \times 5!}{1 \times 5!} = 216,$$

Table 4.1 List of possible groups of 3 out of 6 numbers, if the order of the numbers is not important.

1, 2, 3	1, 3, 4	1, 4, 6	2, 3, 6	3, 4, 5
1, 2, 4	1, 3, 5	1, 5, 6	2, 4, 5	3, 4, 6
1, 2, 5	1, 3, 6	2, 3, 4	2, 4, 6	3, 5, 6
1, 2, 6	1, 4, 5	2, 3, 5	2, 5, 6	4, 5, 6

and the probability is

$$P(\text{winning second prize}) = \frac{\binom{6}{5} \times 36}{\binom{42}{6}} = \frac{216}{5,245,786} \approx 0.000041176.$$

A similar argument applies for the third prize (getting 4 out of 6 numbers). For the number of combinations that match exactly 4 numbers, we need to first choose 4 among the 6 winning numbers, and then 2 numbers among the remaining 36 non-winning numbers. Hence,

$$P(\text{winning third prize}) = \frac{\binom{6}{4}\binom{36}{2}}{\binom{42}{6}} = \frac{9450}{5,245,786} \approx 0.0018014.$$

Finally, for the fourth prize (3 out of 6 numbers) we have

$$P(\text{winning fourth prize}) = \frac{\binom{6}{3}\binom{36}{3}}{\binom{42}{6}} = \frac{142,800}{5,245,786} = 0.027222.$$

The following code can be used to simulate the outcome of the Colorado Lotto and estimate the probability of the third and fourth prizes (see Sidebar 4.3 for how to use R to sample without replacement).

```
> outspc = seq(1,42)
> yourticket = sample(outspc, 6, replace=FALSE)
> n = 200000
> numberofmatches = rep(0, n)
> for(i in 1:n){
+     draw = sample(outspc, 6, replace=FALSE)
+     matches = (draw %in% yourticket)
+     numberofmatches[i] = sum(matches)
+ }
> sum(numberofmatches==4)/n    # Third prize

[1] 0.001855

> sum(numberofmatches==3)/n    # Fourth prize

[1] 0.026755
```

Sidebar 4.3 Sampling Without Replacement in R

The best way to understand sampling with and without replacement is to think about sequentially picking distinct balls from an urn. When we sample with replacement each ball is returned to the urn after being checked. Hence, balls already drawn could potentially show up again in a subsequent try. On the other hand, when sampling without replacement, balls are discarded after they are drawn, so they cannot appear again in the future. Rolling a die is an example of sampling with replacement: drawing a six the first time you roll the die does not prevent you from rolling a six the second time. On the other hand, drawing multiple cards from a deck you are sampling without replacement since the same card cannot appear twice.

In previous chapters, our examples involved only situations in which we were sampling without replacement and we used the function `sample()` with the option `replace = TRUE` to generate random samples. By instead using the option `replace = FALSE`, we can instead run simulations that use sampling without replacement.

```
> sample(seq(1,10), 6, replace=TRUE)   # Numbers can repeat

[1] 3 6 3 8 2 9

> sample(seq(1,10), 6, replace=FALSE) # No repeats

[1] 5 6 4 9 3 7
```

As the following error suggests, the size of the sample cannot be larger than the number of items in the sample space when using sampling without replacement.

```
> sample(seq(1,10), 12, replace=FALSE)

Error in sample.int(length(x), size, replace, prob):
cannot take a sample larger than the population when
'replace = FALSE'
```

4.1.2 The California Superlotto

Let's analyze now the *Superlotto* game offered by the California lottery. In this variant of lotto, 6 numbers are picked; the first 5 are selected between 1 and 47, and the 6th number (called the Mega) is selected between 1 and 27. Note that because the Mega is drawn separately from the other 5 numbers, it might be equal to one of the 5 other numbers. The first prize is awarded to the tickets that match all 6 numbers, other prizes are awarded depending on how many of the first 5 numbers are matched, and on whether the mega is also matched or not.

The number of different tickets in the California Superlotto is

$$\text{Number of different Superlotto tickets} = \binom{47}{5} \times 27 = 41{,}416{,}353,$$

where the first term corresponds to the number of ways in which 5 numbers can be selected out of 47, while the second term corresponds to the number of ways in which the Mega number can be selected. Hence, the probability of winning the first prize is

$$P(\text{winning first prize}) = \frac{1}{41{,}416{,}353}.$$

The second prize in the California Superlotto is awarded to those tickets that match the 5 first winning numbers but do not get the Mega number right. Using the multiplication rule, this number is simply

$P(5 \text{ out of } 5 \text{ but no Mega})$

$$= \frac{\overbrace{1}^{\substack{\text{Numbers of} \\ \text{sets of 5} \\ \text{correct first} \\ \text{numbers}}} \times \overbrace{26}^{\substack{\text{Numbers that} \\ \text{are not the} \\ \text{Mega}}}}{\binom{47}{5} \times 27} = \frac{26}{41{,}416{,}353}.$$

The third prize is awarded to tickets that match the Mega number and 4 out of the 5 first numbers. Using a similar reasoning to the previous examples

$P(4 \text{ out of } 5 \text{ and Mega})$

$$= \frac{\overbrace{\binom{5}{4}}^{\substack{\text{Ways to} \\ \text{choose 4} \\ \text{correct} \\ \text{numbers out} \\ \text{of 5}}} \times \overbrace{42}^{\substack{\text{Ways to} \\ \text{choose 5th} \\ \text{number among} \\ \text{the rest}}} \times \overbrace{1}^{\substack{\text{Ways to} \\ \text{choose the} \\ \text{Mega}}}}{\binom{47}{5} \times 27} = \frac{210}{41{,}416{,}353}.$$

The probability associated with other prizes can be computed in a similar way (e.g., see Exercise 10).

4.2 Sharing Profits: De Méré's Second Problem

Combinatorial numbers can be used to answer another question originally posed to Blaise Pascal by the Chevalier De Méré. This question revolves around how to split the proceeds of the bets when a series of games cannot be completed. For example, assume that John and Monica are betting on the outcome of a series of seven games played between two teams (like the Major League Baseball World Series). Assume also that the two teams are evenly matched (therefore, before they start playing, all possible sequences of seven games are equally likely), that both John and Monica bet $10 on their respective teams, and that the first team to win four games gets its fan the whole pot ($20). After playing four games, Monica's team has won three of them and John's only one. If the series has to be canceled, how should they split the $20 pot?

One possible answer is to split the pot evenly, as if the bets had never been made. However, Monica would (rightfully) argue that, since her team had won more games, she should also get a larger share of the pot. The question is, how much larger should it be?

To answer this question, we first need to compute the probability that Monica's team wins its fourth game before John's wins two more. Since the space is equiprobable, the exact history of how we got to the current state does not really matter (i.e., it does not matter who won what during the first four games, as long as we have three wins and one loss for Monica's team). Thus, we could say that the history is,

L W W W _ _ _

where W means that Monica's team won, L means that it lost, and the underlined spaces correspond to the unknown outcomes associated with the last three games. Now, let's consider the future. Since we would typically stop playing once one of the teams has reached four wins, there are four possible ways in which the World Series could end up being played.

History	Winner
L W W W W	Monica
L W W W L W	Monica
L W W W L L W	Monica
L W W W L L L	John

Now, it is tempting to argue that, since three out of those four futures lead to Monica winning the bet, then the probability of her team winning is 3/4. However, this is not quite right because the four outcomes above are not equiprobable. Indeed, it is the whole sequences of seven characters which are equiprobable! Accordingly, we need to consider all the possible sequences of seven characters that start with L W W W (there are 8 of them):

History	Winner
L W W W W W W	Monica
L W W W W L W	Monica
L W W W W W L	Monica
L W W W W L L	Monica
L W W W L W W	Monica
L W W W L W L	Monica
L W W W L L W	Monica
L W W W L L L	John

Note that the first seven imply that Monica wins the bet (the first four correspond to Monica's team winning the fifth game in the series, the next two correspond to Monica's team losing the fifth but winning the sixth, and the second to last corresponds to Monica's team losing the fifth and sixth games but winning the seventh), while only the last one implies that Monica will lose the bet. Therefore, her probability of winning is $7/8 = 0.875$ and not $3/4 = 0.75$!

As before, we can convince ourselves that this reasoning is correct using a simple simulation:

```
> n = 10000
> outspc = c("W", "L")   # From Monica's perspective
> gameres = matrix(0, nrow=n, ncol=3)
> for(i in 1:n){
+    gameres[i,] = sample(outspc, 3, replace=TRUE)
+ }
> numwins = rowSums(gameres=="W")
> sum(numwins >= 1)/n    # Monica needs one or more wins

[1] 0.8724
```

Once we have computed the probability that Monica will win, we can go back to our definition of a fair game and compute Monica's share of the pot as Monica's expected profit:

$$E(\text{payout for Monica}) = 20 \times \frac{7}{8} + 0 \times \frac{1}{8} = 17.5,$$

while John's share should be

$$E(\text{payout for John}) = 20 \times \frac{1}{8} + 0 \times \frac{7}{8} = 2.5$$

(note that both add up to $20, as they should).

This result can be generalized. Suppose that in the current state of the game John needs to win n games to win the bet, and Monica needs to win m of them. In our example above $n = 3$ and $m = 1$. Then, we need to consider an additional

$n + m - 1$ rounds of the game (in our example, we considered $3 + 1 - 1 = 3$). There are 2^{n+m-1} possible different outcomes for these $n + m - 1$ rounds ($2^3 = 8$ in our example), of which

$$\binom{n+m-1}{m} + \binom{n+m-1}{m+1} + \cdots + \binom{n+m-1}{n+m-1}$$

have Monica as the winner. In the previous sum, the first terms correspond to the number of future sequences of $n + m - 1$ games in which Monica wins exactly m games (the minimum it needs to get the full pot), the second corresponds to the number of sequences in which she wins exactly $m + 1$ games, and so on.

From the previous results, we can get the probability that Monica will win the bet is simply

$$P(\text{Monica wins the bet}) = \frac{\binom{n+m-1}{m} + \binom{n+m-1}{m+1} + \cdots + \binom{n+m-1}{n+m-1}}{2^{n+m-1}}.$$

Incidentally, note that if $n = m$ (i.e., both teams are tied at the time the series is halted) then

$$P(\text{Monica wins the bet}) = \frac{1}{2},$$

which implies that the pot should be evenly split (as we would have expected, given that the teams are evenly matched).

4.3   Exercises

1. You are a photographer sitting in a group of 10 people in a row for pictures. How many different seating arrangements could you use?

2. Eight horses (Alabaster, Beauty, Candy, Doughty, Excellente, Friday, Great One, and High 'n Mighty) run a race. In how many ways can the first three finishers turn out?

3. A statistics class has 30 students. The students need to select a team of 5 people to represent them, how many different such teams can be formed?

4. In how many ways can I seat 5 people in a circular table?

5. In casino keno, a player chooses 10 numbers out of 80. If she matches all the 10 numbers, she wins the first prize. What is the probability of winning the first prize in this game? How does it compare with the probability of

winning the Colorado Lotto (use an odds ratio to compare the results, and interpret it)?

6. The Florida Lotto is a lottery offered in the state of Florida; the first prize is won when you get 6 winning numbers from a list of 53 numbers. What is the probability of winning the first prize? The second prize is won if you get 5 of those 6 winning numbers, what is the probability of winning the second prize?

7. What is the probability of winning the first prize of the Florida Lotto if you buy 100 tickets? (Assume each ticket has a different set of numbers.)

8. For the New York State lottery the first prize is won if you get 6 winning numbers from a list of 59 numbers. Calculate the probability of winning this lottery. The third prize is obtained by getting 4 of the 6 winning numbers, calculate also the probability for this prize.

9. What is the probability of winning this last lottery if you buy 100 tickets? What is your expected profit for this situation if each ticket costs $1?

10. What is the probability of winning the fourth prize in the California Superlotto? The fourth prize goes to tickets that get 4 out of the 5 first numbers correct, but miss the Mega.

11. Imagine the California SuperLotto lottery is changed so that there's a second Mega number (chosen from the same list of 26 numbers as the first Mega number) and the first prize is obtained if the player gets the 5 winning numbers, the first Mega number and the second Mega number. What is the probability of getting the first prize in the new lottery? Is this prize harder or easier to win than the actual California SuperLotto?

12. In the California Lotto, what is the probability of getting any 3 of the 5 winning numbers and the Mega?

13. In the SuperLotto, you can get a prize if you get 4 out of the 5 winning numbers and the Mega number, but you also get a price if you just get 4 out of the 5 winning numbers and miss the Mega number. Which of the two prizes has higher probability of winning?

14. Which is more likely: to get 3 out of the 5 winning numbers and the Mega number or getting the 4 out of 5 winning numbers?

15. [R] Can you list out all the combinations of 3 numbers from the list of numbers from 1 to 6? Check your list using R.

16. [R] Modify the R code for simulating the Colorado Lotto in order to instead estimate the probability of the third and fourth prizes of the California Superlotto.

5

The Monty Hall Paradox and Conditional Probabilities

The Monty Hall problem is loosely based on the American television game show *Let's Make a Deal* and it is named after the show's original host, Monty Hall. It is considered a paradox because the result appears absurd but can be demonstrated to be true nevertheless. The problem was made famous when it appeared in Marilyn von Savant's *Parade Magazine* column in 1990, and it has also been featured in the final episode of the 2004–2005 season of *NUMB3RS*, as well as in the opening scenes of the 2008 movie *21*.

5.1 The Monty Hall Paradox

Suppose you are participating in a contest and you are given a choice of three doors: behind one door there is a car; behind the others, goats. You pick a door, say number 1, and the host (call him *Monty*), who knows what's behind each door, opens another door, say number 3, which has a goat. He then offers you the opportunity to switch to the other unopened door (in this case, door number 2). Should you switch doors?

Contrary to intuition, under some reasonable assumptions (mainly, that when a random choice needs to be made, all options are chosen with the same probability), you are better off switching because your probability of winning the car increases. To see why this is true, we will represent the contest as a series of decisions using a *tree diagram*. In a tree diagram, each level of the tree represents a series of mutually exclusive events that can occur at a given point in time. By following the different branches of the tree from its root, we can represent all possible outcomes in a complex experiment.

The Monty Hall paradox involves three different sets of events: the producers of the contest need to decide which doors hides the car, the contestant (you) needs to pick a door, and finally Monty needs to decide which other

Probability, Decisions and Games: A Gentle Introduction using R, First Edition. Abel Rodríguez and Bruno Mendes.
© 2018 John Wiley & Sons, Inc. Published 2018 by John Wiley & Sons, Inc.
Companion website: www.wiley.com/go/Rodriguez/Probability_Decisions_and_Games

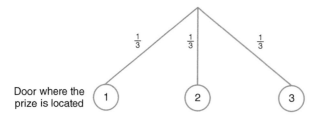

Door where the
prize is located

Figure 5.1 Each branch in this tree represents a different decision and the 1/3s represent the probability of each door being selected to contain the prize.

(among the ones that hide a goat and have not been chosen by you) he will open. Hence, our final tree will consist of three levels, which we will be adding to it sequentially.

Consider the first choice. Before the contest starts, the producers are free to place the car behind any one of the three doors. Since we assume that each door is selected with the same probability, we end up with the tree representation in Figure 5.1. The number next to each branch corresponds to the probability we associate with each possible outcome. In this case, the probability of each branch is 1/3 because we assumed that the prize is located behind each door with the same probability.

Now, for the second decision, you are unaware of which door hides the car, so you are also free to select any of the three doors. Again, we assume that you select each door with the same probability, leading to the representation in Figure 5.2.

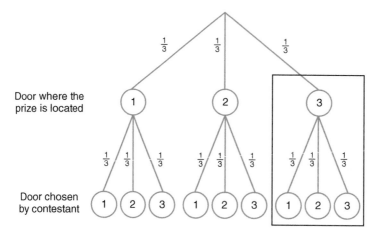

Door where the
prize is located

Door chosen
by contestant

Figure 5.2 The tree structure now represents an extra level, representing the contestant decisions and the probability for each decision to be the one chosen.

As a consequence of the multiplication rule, if we follow along the branches of the tree, we can obtain the probability associated with any combination of doors by multiplying together the corresponding probabilities. Therefore, in this case, any combination has the same probability 1/9.

Consider now Monty's decision: unlike the previous ones, his options are affected by the choices made by the producers and you, the contestant. To simplify the explanation, consider only the branch corresponding to the car being located behind door 3. Then, if the door chosen by you is either door 2 or door 3, Monty has a single choice for the door he will open (he cannot open the door with the car, or the door chosen by you). On the other hand, if you chose door 1, Monty has two options (he can open either door 2 or door 3), and from our original description, he opens each door with the same probability. Figure 5.3 shows the sub-tree associated with these decisions.

A similar argument applies if the prize is located behind either door 2 or door 3. This leads to 12 possible paths with non-zero probability, as shown in Table 5.1 (note that the sum of all probabilities equals 1). Of these 12 paths, six (the ones highlighted in the table) correspond to paths where you would win by switching. If we sum these six values (the branches correspond to mutually exclusive events), we get

$$P(\text{winning the car if you switch}) = \frac{1}{9} + \frac{1}{9} + \frac{1}{9} + \frac{1}{9} + \frac{1}{9} + \frac{1}{9} = \frac{2}{3},$$

which shows that it is beneficial for the player to switch doors.

You can empirically verify these results by running the following simulation in R:

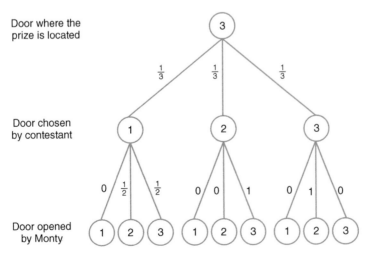

Figure 5.3 Decision tree for the point when Monty decides which door to open assuming the prize is behind door 3.

Table 5.1 Probabilities of winning if the contestant in the Monty problem switches doors.

Door where the prize is	Door chosen by contestant	Door opened by Monty	Probability of the scenario
1	1	2	1/18
1	1	3	1/18
1	2	3	1/9
1	3	2	1/9
2	1	3	1/9
2	2	1	1/18
2	2	3	1/18
2	3	1	1/9
3	1	2	1/9
3	2	1	1/9
3	3	1	1/18
3	3	2	1/18

Highlighted lines correspond to scenarios in which the player wins by switching doors.

```
> door = seq(1,3)
> n = 10000
> winifswitch = rep(FALSE, n)
> for(i in 1:n){
+    prizelocation = sample(door, 1)
+    contestantchoice = sample(door, 1)
+    if(prizelocation==contestantchoice){
+      dooropened = sample(door[-contestantchoice], 1)
+    }else{
+      dooropened = door[-c(prizelocation,contestantchoice)]
+    }
+    doorifswitch = door[-c(dooropened,contestantchoice)]
+    winifswitch[i] = (doorifswitch==prizelocation)
+ }
> sum(winifswitch)/n

[1] 0.6707
```

5.2 Conditional Probabilities

The solution to the Monty Hall problem illustrates the concept of *conditional probability*. Conditional probabilities simply reflect the fact that the probability of an event might depend on our knowledge of whether other events have already occurred in the past or not. For example, the probability of winning in the Monty Hall problem (i.e., choosing the door that hides the car) before Monty opens the door is 1/3. However, once Monty has opened a door, the probability of winning changes. This happens because we have learned something about the space of outcomes from the fact that Monty opened a door.

Conditional probabilities involve two events; in our example above we have

$$A = \{\text{Win the car}\} \quad \text{and} \quad B = \{\text{Switching door}\},$$

and we are interested in computing the probability of A if B has happened (i.e., the probability that you win the car if you switch doors), which we will denote $P(A \mid B)$. This expression is also read as "the probability of A given B" or "the probability of A conditional on B." Unlike the event whose probability we want to compute (in this case, A, or winning the car), the event we condition upon (in this case, B, switching doors) is *not* random; B is an event we assume has occurred. Consequently, generally speaking $P(A \mid B) \neq P(B \mid A)$. As a matter of fact, it is easy to confuse the conditional probability of one event given another with their joint probability of these events. Recall that the joint probability of A and B, denoted $P(A \cap B)$, describes the probability that A and B happen *simultaneously*. In this case, both events are random (we do not know if they have occurred or not), and we have that $P(A \cap B) = P(B \cap A)$.

To further illustrate the difference between joint and conditional probabilities, consider examining the association between smoking and lung cancer. More specifically, let's imagine we interview 1000 persons and ask them whether they have ever smoked and also whether they have suffered from lung cancer (the results of one such study are presented in Table 5.2). Let

$$S = \{\text{A randomly chosen person smokes}\}$$
$$C = \{\text{A randomly chosen person has suffered from lung cancer.}\}$$

We could ask what the probability is that a randomly selected person smokes *and* suffers from lung cancer. If we do this, we are inquiring about the joint probability of S and C, $P(S \cap C)$. By exploiting the frequentist interpretation of probability, we could estimate $P(S \cap C)$ by dividing the number of people

Table 5.2 Studying the relationship between smoking and lung cancer.

	Had lung cancer	Never had lung cancer
Has smoked	20	180
Has not smoked	50	750

who answered both questions affirmatively by the total number of people interviewed. Hence,

$$P(S \cap C) = \frac{20}{20 + 180 + 50 + 750} = 0.02.$$

As we can see, that probability is small. On the other hand, we could also ask what is the probability of suffering lung cancer if you smoke, that is, $P(C \mid S)$. Clearly, this is the most relevant question if you want to decide if you want to quit smoking or not. Note that in this case there is nothing random about whether the person smokes or not: we know that he or she does. So, we need to compute $P(C \mid S)$, and we need to look only at the people who have suffered from cancer among those that smoke, that is,

$$P(C \mid S) = \frac{20}{20 + 180} = 0.1.$$

Note how different $P(S \cap C)$ and $P(C \mid S)$ are. Furthermore, if we define $\bar{S} = \{A$ randomly chosen person does *not* smoke$\}$, we could compute $P(C \mid \bar{S})$, the probability of suffering from cancer if you do not smoke,

$$P(C \mid \bar{S}) = \frac{50}{50 + 750} = 0.0625.$$

All these calculations indicate that smokers are about one and a half times more likely than nonsmokers to suffer from lung cancer (since $P(C \mid S)/P(C \mid \bar{S}) = 1.6$). Furthermore, note that $P(C \mid S) + P(C \mid \bar{S}) \neq 1$, but $P(C \mid S) + P(\bar{C} \mid S) = 1$.

Even though joint and conditional probabilities are different concepts, there is a link between them,

> The conditional probability of an event A given B can be computed as
> $$P(A \mid B) = \frac{P(A \cap B)}{P(B)},$$
> or alternatively
> $$P(A \cap B) = P(A \mid B)P(B).$$

We implicitly used the second formula while constructing the decision tree for the *Monty Hall* problem. In fact, the probabilities in the branches of the tree are all conditional probabilities given the previous events in the tree. For example, if we define the events

$A_1 = \{$Producers put the car behind door 1$\}$,

$B_1 = \{$Contestant choses door 1$\}$,

$C_2 = \{$Monty opens door 2.$\}$

Then the very first branch represents $P(A_1) = 1/3$, while $P(B_1 \mid A_1) = 1/3$ and $P(C_2 \mid B_1 \cap A_1) = 1/2$. Using these values, we computed the joint probability of all events in the branch as the product of all three values (just as the formula suggests), that is,

$$P(A_1 \cap B_1 \cap C_2) = P(C_2 \mid B_1 \cap A_1)P(B_1 \mid A_1)P(A_1).$$

5.3 Independent Events

In the first few chapters, and in particular when discussing roulette, we informally used the term *independent* to qualify experiments where the outcome of one trial does not affect the outcome of another. Conditional probabilities can be used to formalize the notion of independence.

Independent Events

We say that two events A and B are independent if

$$P(A \mid B) = P(A),$$

which implies that

$$P(A \cap B) = P(A)P(B).$$

Intuitively, two events are independent if knowledge of whether B happened or not does not affect the probability of A happening. To clarify this notion, consider the Monty Hall problem again. In that case, the selection of the door by the contestant is independent from the selection by the producers. However, the decision by Monty is not independent from the decisions made by the producers and the contestant, which is the reason for the apparent paradox.

In our example involving smoking and lung cancer, we can see that cancer and smoking are not independent: since $P(C \cap S) = 0.02$ and also $P(C) = \frac{20+50}{1000} = 0.07$ and $P(S) = \frac{20+180}{1000} = 0.2$, then $P(C)P(S) = 0.014 \neq P(C \cap S)$.

One consequence of independence is that formulas for the expectation and variance of random variables simplify

---

If X and Y are independent random variables and a and b are two constant (non-random numbers), then

$$E(XY) = E(X)E(Y)$$

$$V(aX + bY) = a^2 V(X) + b^2 V(Y)$$

5.4 Bayes Theorem

As we discussed before, conditional probabilities are not symmetric, that is, in general $P(A \mid B) \neq P(B \mid A)$. However, the two quantities are related. Indeed, since $P(A \cap B) = P(A \mid B)P(B)$ and also $P(A \cap B) = P(B \mid A)P(A)$, we have that

---

Bayes Theorem
For any two events A and B,

$$P(A \mid B) = \frac{P(B \mid A)P(A)}{P(B)}$$

When $P(B)$ is unknown we can compute it from $P(B \mid A)$, $P(B \mid \overline{A})$, $P(A)$, and $P(\overline{A})$. Figure 5.4 illustrates how an event B can be broken down in two parts, $B \cap A$ and $B \cap \overline{A}$. By exploiting this decomposition, we obtain the following representation:

Total Probability Law
For any two events A and B,

$$P(B) = P(B \cap A) + P(B \cap \overline{A}) = P(B \mid A)P(A) + P(B \mid \overline{A})P(\overline{A})$$

Substituting back into Bayes theorem, we have

Bayes Theorem (Alternative Formulation)
For any two events A and B,

$$P(A \mid B) = \frac{P(B \mid A)P(A)}{P(B)} = \frac{P(B \mid A)P(A)}{P(B \mid A)P(A) + P(B \mid \overline{A})P(\overline{A})}$$

A slight generalization of Bayes theorem involves a more general partition of the space with more than two choices.

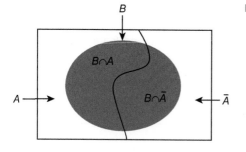

Figure 5.4 Partitioning the event space.

General Alternative Formulation

For any events B and $A_1, A_2, \ldots, A_k$ such that $A_1, A_2, \ldots, A_k$ form a partition of the space of all possible outcomes,

$$P(A_i \mid B) = \frac{P(B \mid A_i)P(A_i)}{P(B)}$$

$$= \frac{P(B \mid A_i)P(A_i)}{P(B \mid A_1)P(A_1) + \cdots + P(B \mid A_k)P(A_k)}$$

Bayes theorem is particularly useful because in many cases we might be interested in $P(A \mid B)$, but it is much easier to compute $P(B \mid A)$. For example, suppose that you play the following game, which we call the *game of urns*. There are three urns:

- Urn 1 contains 2 blue balls, 2 red balls and 2 yellow balls.
- Urn 2 contains 3 blue balls and 2 red balls.
- Urn 3 contains 2 blue balls and 4 yellow balls.

The dealer secretly picks an urn uniformly at random, from which she draws a ball also uniformly at random. The dealer then presents the ball to you and asks you to decide what urn the ball was drawn from. If you pick the right urn, you get your wager back plus \$1. Otherwise, you lose the money you bet. The question is, how should you play?

Clearly, your answer should depend on what color shows up. For example, you do not need to be an expert in probability to realize that if you observe a yellow ball you should never pick urn 2 (it does not contain any yellow ball in the first place!). However, intuition is less useful in deciding whether to pick urn 1 or urn 3.

We can use conditional probabilities and Bayes theorem to generate a strategy. Let

$$Y = \{\text{Ball is yellow}\}, \qquad U_1 = \{\text{Ball extracted from urn 1}\},$$

$$R = \{\text{Ball is red}\}, \qquad U_2 = \{\text{Ball extracted from urn 2}\},$$

$$B = \{\text{Ball is blue}\}, \qquad U_3 = \{\text{Ball extracted from urn 3}\}.$$

Consider first the case when the dealer shows you a yellow ball. To create a strategy, we need to compute the probabilities associated with the ball coming from each of the urns conditional on it being yellow, that is, $P(U_1 \mid Y)$, $P(U_2 \mid Y)$, and $P(U_3 \mid Y)$. Then, the optimal strategy is to select the urn with the highest probability. Using Bayes theorem, we have

$$P(U_1 \mid Y) = \frac{P(Y \mid U_1)P(U_1)}{P(Y \mid U_1)P(U_1) + P(Y \mid U_2)P(U_2) + P(Y \mid U_3)P(U_3)},$$

$$P(U_2 \mid Y) = \frac{P(Y \mid U_2)P(U_2)}{P(Y \mid U_1)P(U_1) + P(Y \mid U_2)P(U_2) + P(Y \mid U_3)P(U_3)},$$

$$P(U_3 \mid Y) = \frac{P(Y \mid U_3)P(U_3)}{P(Y \mid U_1)P(U_1) + P(Y \mid U_2)P(U_2) + P(Y \mid U_3)P(U_3)},$$

(note that the denominator is the same for all three expressions, and that it corresponds to $P(Y)$ because the three events U_1, U_2, and U_3 form a partition of all possible events).

Since the dealer picks the urns uniformly at random, we have $P(U_1) = P(U_2) = P(U_3) = 1/3$. Also, $P(Y \mid U_1) = 2/6 = 1/3$ (because urn 1 has two yellow balls out of six), $P(Y \mid U_2) = 0$ (because there are no yellow balls in urn 2), and $P(Y \mid U_3) = 4/6 = 2/3$ (because urn 3 has four yellow balls out of six). Substituting these values we have

$$P(U_1 \mid Y) = \frac{\frac{1}{3} \times \frac{1}{3}}{\frac{1}{3} \times \frac{1}{3} + 0 \times \frac{1}{3} + \frac{2}{3} \times \frac{1}{3}} = \frac{1}{3},$$

$$P(U_2 \mid Y) = \frac{0 \times \frac{1}{3}}{\frac{1}{3} \times \frac{1}{3} + 0 \times \frac{1}{3} + \frac{2}{3} \times \frac{1}{3}} = 0,$$

$$P(U_3 \mid Y) = \frac{\frac{2}{3} \times \frac{1}{3}}{\frac{1}{3} \times \frac{1}{3} + 0 \times \frac{1}{3} + \frac{2}{3} \times \frac{1}{3}} = \frac{2}{3}.$$

Therefore, if we see a yellow ball, the optimal strategy is to select urn 3. Incidentally, note our calculation gives us the probability of drawing a yellow ball as a byproduct, the probability of a yellow ball is $P(Y) = \frac{1}{3} \times \frac{1}{3} + 0 \times \frac{1}{3} + \frac{2}{3} \times \frac{1}{3} = \frac{1}{3}$.

A similar approach can be used in the case we observe a red ball,

$$P(U_1 \mid R) = \frac{\frac{1}{3} \times \frac{1}{3}}{\frac{1}{3} \times \frac{1}{3} + \frac{2}{5} \times \frac{1}{3} + 0 \times \frac{1}{3}} = \frac{5}{11},$$

$$P(U_2 \mid R) = \frac{\frac{2}{5} \times \frac{1}{3}}{\frac{1}{3} \times \frac{1}{3} + \frac{2}{5} \times \frac{1}{3} + 0 \times \frac{1}{3}} = \frac{6}{11},$$

$$P(U_3 \mid R) = \frac{0 \times \frac{1}{3}}{\frac{1}{3} \times \frac{1}{3} + \frac{2}{5} \times \frac{1}{3} + 0 \times \frac{1}{3}} = 0,$$

where, as a byproduct, we see that $P(R) = \frac{1}{3} \times \frac{1}{3} + \frac{2}{5} \times \frac{1}{3} + 0 \times \frac{1}{3} = \frac{11}{45}$. This means that the optimal strategy in this case is to attribute the red ball to urn 2.

Finally, for a blue ball

$$P(U_1 \mid B) = \frac{\frac{1}{3} \times \frac{1}{3}}{\frac{1}{3} \times \frac{1}{3} + \frac{3}{5} \times \frac{1}{3} + \frac{1}{3} \times \frac{1}{3}} = \frac{15}{57},$$

$$P(U_2 \mid B) = \frac{\frac{3}{5} \times \frac{1}{3}}{\frac{1}{3} \times \frac{1}{3} + \frac{2}{5} \times \frac{1}{3} + \frac{1}{3} \times \frac{1}{3}} = \frac{9}{19},$$

$$P(U_3 \mid B) = \frac{\frac{1}{3} \times \frac{1}{3}}{\frac{1}{3} \times \frac{1}{3} + \frac{2}{5} \times \frac{1}{3} + \frac{1}{3} \times \frac{1}{3}} = \frac{15}{57},$$

while $P(B) = \frac{1}{3} \times \frac{1}{3} + \frac{3}{5} \times \frac{1}{3} + \frac{1}{3} \times \frac{1}{3} = \frac{19}{45}$, so the optimal strategy is again to attribute the blue ball to urn 2.

The probability of winning this game under the optimal strategy can be obtained using the total probability law. Figure 5.5 shows a tree representation

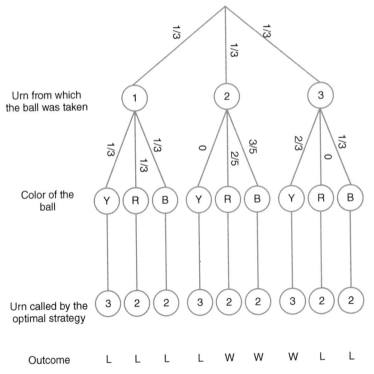

Figure 5.5 Tree representation of the outcomes of the *game of urns* under the optimal strategy that calls yellow balls as coming from Urn 3 and blue and red balls as coming from urn 2.

of the game. Note that the only ways to win are when a blue ball comes up from urn 2 (which is correctly called by our strategy) and when a yellow ball is taken from the third urn (which is again correctly called by our strategy). Hence,

$$P(\text{Win}) = P(B|U_2)P(U_2) + P(R|U_2)P(U_2) + P(Y|U_3)P(U_3)$$

$$= \frac{3}{5} \times \frac{1}{3} + \frac{2}{5} \times \frac{1}{3} + \frac{4}{6} \times \frac{1}{3} = \frac{50}{90} \approx 0.5556.$$

The following R code simulates the game of urns and can be used to check that the derivations shown above are correct.

```
> n = 10000
> urnspc = seq(1,3)
> colorpr = matrix(c(1/3,1/3,1/3,
+                    0,  2/5,3/5,
+                    2/3,0,  1/3), nrow=3, ncol=3, byrow=T)
> balls = c("Y", "R", "B")
> colnames(colorpr) = balls
> urn          = rep(0,n)
> ballcolor    = rep(0,n)
> optimalcall = rep(0,n)
> for(i in 1:n){
+    urn[i] = sample(urnspc,1)
+    ballcolor[i] = sample(balls, 1, replac=T,
+                          prob=colorpr[urn[i],])
+    if(ballcolor[i]=="Y"){
+      optimalcall[i] = 3
+    }else{
+      optimalcall[i] = 2
+    }
+ }
> sum(ballcolor=="Y")/n  # Prob of a Yellow Ball

[1] 0.3328

> sum(ballcolor=="B")/n  # Prob of a Blue Ball

[1] 0.4246

> sum(optimalcall==urn)/n  # Prob of winning with opt strat

[1] 0.5579
```

5.5 Exercises

1. There are three condemned prisoners in jail, one of whom is to be secretly pardoned. One of the prisoners begs the warden to tell him the

name of one of the others who will be executed, arguing that this reveals no information about his own fate but increases his chances of being pardoned from 1/3 to 1/2. The warden obliges, (secretly) flipping a coin to decide which name to provide if the prisoner who is asking is the one being pardoned. Does knowing the warden's answer really change the asking prisoner's chances of being pardoned?

2. *Ignorant Monty*: In a variant of the Monty Hall problem, Monty does not know what lies behind the door and picks one at random to open. When he does, he is relieved that it contains a goat. Show that, in this case, it is irrelevant whether you switch or not.

3. **[R]** Write a simulation that corroborates your calculations for the *ignorant Monty* game.

4. You are presented with three boxes: a box containing two gold coins, a box with two silver coins, and a box with one of each. After choosing a box at random and withdrawing one coin at random that happens to be a gold coin, what is the probability that the other coin is gold?

5. Consider the following table giving the joint probability for two random variables. Are the two events $X = 2$ and $Y = 5$ independent? How about $X = 2$ and $Y = 1$?

	$Y = 1$	$Y = 5$
$X = 2$	0.25	0.25
$X = 5$	0.25	0.25

6. Consider the following table giving the joint probability for two random variables. Are the two events $X = 0$ and $Y = 0$ independent? How about $X = 0$ and $Y = 2$?

	$Y = 0$	$Y = 1$	$Y = 2$
$X = 0$	0.25	0.15	0.10
$X = 1$	0.25	0.20	0.05

7. When a laboratory tests you for a particular condition (say the test is for the presence of human immunodeficiency virus, or HIV) it can either produce a positive or negative result. The latter means there is no virus in your

body, the former means there is a virus in your system. These tests have rates of *sensitivity* (i.e., how often they correctly diagnose a person with the disease) and rates of *specificity* (i.e., the rate of times the test correctly identifies people who do not have the condition). These rates are ideally close to 100% but, in practice, there are always false positives and false negatives. In other words, there are always situations in which the test says someone has the virus but the person actually doesn't, and situations where the test says there is no virus in the person's system but there actually is. Let's say we are testing a new HIV test and the results are presented in the following 2×2 table.

	Patients with HIV	Patients with no HIV
Patient with positive test	10	1
Patient with negative test	10	10,000

Based on the data and using a frequentist approach to probability, answer the following questions
(a) What's the probability of a person having HIV?
(b) What is the probability of the test correctly diagnosing the presence of HIV (i.e., what is the sensitivity of the test)?
(c) What is the specificity of the test?
(d) What is the rate of false positives?
(e) What is the rate of false negatives?
(f) In terms of the usefulness to the society at large (remember this is a communicable disease), what is more useful: a lower rate of false positives or a lower rate of false negatives?

8. Consider a competitor HIV test to the one mentioned above. Its field tests produced the following data.

	Patients with HIV	Patients with no HIV
Patient with positive test	9	2
Patient with negative test	5	10,000

Based on the data and using a frequentist approach to probability, answer the following questions
(a) Calculate P(test is positive | person has HIV)?
(b) Calculate P(test is negative | person doesn't have HIV)?

(c) Calculate P(test is positive | person doesn't have HIV)?

(d) Calculate P(test is negative | person has HIV)?

(e) Which of the two tests is preferable?

9. Compute the fair value of the *game of urns* described at the end of the chapter.

10. Consider a variation of the game of urns where
   - Urn 1 contains 4 blue balls, 2 red balls, and 1 yellow ball.
   - Urn 2 contains 1 blue ball, 2 red balls, and 2 yellow balls.
   - Urn 3 contains 2 blue balls, 1 blue ball, and 3 yellow balls.

   What would your optimal strategy be in this case, and what is the fair value of this game?

11. **[R]** Write a simulation for the game of urns in the previous exercise.

6

Craps

Craps is the most popular dice game in casinos. The game has been featured in multiple movies including *Ocean's Thirteen* (2007), *Snake Eyes* (1998), and *Big Town* (1987). The mathematical analysis of the game of craps is similar in some ways to that of roulette (both games involve independent rounds of play), but because each round is composed of 2 interdependent phases that have different rules, the analysis has to be carried out carefully.

6.1 Rules and Bets

In craps you are betting on the outcome of two dice rolled simultaneously. An appealing feature of the game is that you can play it either as the *shooter* (if you are the one rolling the dice) or as a *stand-by* (if you are a spectator, by betting with or against the shooter). As with roulette, players place their bets by placing their chips on the appropriate sections of the board (see Figure 6.1). The nicknames associated with each of the outcomes are presented in Table 6.1.

Each round of craps is comprised of two phases. The first phase consists of a single roll called the *come-out* roll, and the second phase (which might consist of multiple rolls) is called the *point*.

6.1.1 The Pass Line Bet

The *pass line* (also called *win* or *right*) bet is the most basic bet in craps, and the shooter is obligated to make this wager in order to play. In addition to the shooter, any spectator can participate in the pass line bet. Typically, the pass line bet pays even odds (recall that this means that, if you win, you get your money back plus a profit equal to your bet).

The outcome of the pass line bet is resolved as follows. If the come-out roll is a 7 or 11 (called a *natural*), then the pass line bet wins automatically, and the round ends (there is no second phase in that case). Similarly, if the come-out roll is a 2, 3, or 12, then the pass line bet loses automatically, and again the round

Probability, Decisions and Games: A Gentle Introduction using R, First Edition. Abel Rodríguez and Bruno Mendes.
© 2018 John Wiley & Sons, Inc. Published 2018 by John Wiley & Sons, Inc.
Companion website: www.wiley.com/go/Rodriguez/Probability_Decisions_and_Games

Table 6.1 Names associated with different combinations of dice in craps.

	1	2	3	4	5	6
1	Snake eyes	–	–	–	–	–
2	Ace ddeuce	Hard four	–	–	–	–
3	Easy four	Five (fever five)	Hard six	–	–	–
4	Five (fever five)	Easy six	Natural or seven out	Hard eight	–	–
5	Easy six	Natural or seven out	Easy eight	Nine (nina)	Hard ten	–
6	Natural or seven out	Easy eight	Nine (nina)	Easy ten	Yo (Yo-leven)	Boxcars or Midnight

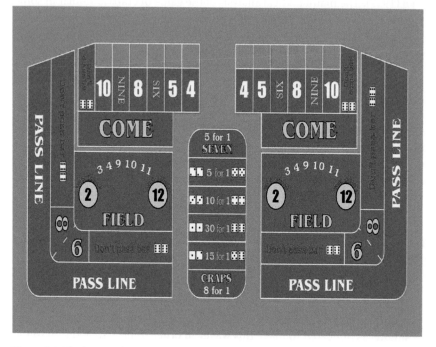

Figure 6.1 The layout of a craps table.

ends. Losing in this way is often referred to as *crapping out*. Finally, if any other number is rolled (i.e., a 4, 5, 6, 8, 9, or 10), that number becomes the point and we move into the second phase of the round.

When a point is established, the goal of the game changes. The shooter keeps rolling the dice until either the point comes up again or a 7 comes out. If the

Table 6.2 All possible equiprobable outcomes associated
with two dice being rolled.

1–1	2–1	3–1	4–1	5–1	6–1
1–2	2–2	3–2	4–2	5–2	6–2
1–3	2–3	3–3	4–3	5–3	6–3
1–4	2–4	3–4	4–4	5–4	6–4
1–5	2–5	3–5	4–5	5–5	6–5
1–6	2–6	3–6	4–6	5–6	6–6

Table 6.3 Sum of points associated with the roll of two dice.

Sum of the dice	List of ways the sum can be realized	Number of ways the sum can be realized
2	1–1	1
3	1–2, 2–1	2
4	1–3, 3–1, 2–2	3
5	1–4, 4–1, 2–3, 3–2	4
6	1–5, 5–1, 2–4, 4–2, 3–3	5
7	1–6, 6–1, 2–5, 5–2, 3–4, 4–3	6
8	2–6, 6–2, 3–5, 5–3, 4–4	5
9	3–6, 6–3, 4–5, 5–4	4
10	4–6, 6–4, 5–5	3
11	5–6, 6–5	2
12	6–6	1

point comes out first, then the pass line bets win. On the other hand, if a 7 comes out first (referred to as *seven out*), then the pass line bets lose. Note that this is the opposite of what happens in the come-out roll, where a 7 wins the game.

Let's analyze the pass line bet and compute the house advantage in craps. First, recall that there are 36 outcomes for the roll of two dice, and that, as long as the dice are fair, they are all equiprobable (see Table 6.2). To analyze craps, it is convenient to group these 36 outcomes into 11 groups, depending on what their sum is (see Table 6.3).

As we did for the Monty Hall problem, we can use a tree to help us compute the probability of winning at craps (see Figure 6.2). We proceed now to fill in the probabilities associated with each of the branches in the tree.

Since all 36 outcomes are equiprobable, it is easy to see that the probability of getting a natural in the come-out roll (i.e., winning in the first phase if you

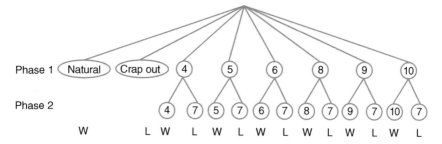

Figure 6.2 Tree representation for the possible results of the game of craps. Outcomes that lead to the pass line bet winning are marked with W, while those that lead to a lose are marked L.

make the pass line bet) is simply equal to

$$P\left(\begin{array}{c}\text{winning the pass line bet}\\ \text{in the first phase}\end{array}\right) = P(\text{getting a 7 } \underline{or} \text{ getting an 11})$$

$$= P(\text{getting a 7}) + P(\text{getting an 11})$$

$$= \frac{6+2}{36} = \frac{8}{36} = \frac{2}{9}.$$

Similarly, we can compute the probability of crapping out

$$P(\text{crapping out}) = P(\text{getting a 2 } \underline{or} \text{ getting a 3 } \underline{or} \text{ getting a 12})$$

$$= P(\text{getting a 2}) + P(\text{getting a 3}) + P(\text{getting a 12})$$

$$= \frac{1+2+1}{36} = \frac{4}{36} = \frac{1}{9}.$$

and the probability of getting each of the points is

$$P(\text{point is 4}) = \frac{3}{36} = \frac{1}{12},$$

$$P(\text{point is 5}) = \frac{4}{36} = \frac{1}{9},$$

$$P(\text{point is 6}) = \frac{5}{36},$$

$$P(\text{point is 8}) = \frac{5}{36},$$

$$P(\text{point is 9}) = \frac{4}{36} = \frac{1}{9},$$

$$P(\text{point is 10}) = \frac{3}{36} = \frac{1}{12}.$$

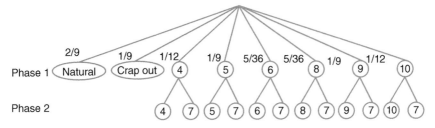

Figure 6.3 Tree representation for the possible results of the game of craps with the probabilities for each of the come-out roll.

Filling these numbers into the tree we obtain Figure 6.3. Note that these probabilities sum to 1, as we would have expected. Also, note that the probability of winning during the come-out roll (2/9) is much larger than the probability of losing during the come-out roll (1/9), and that both are much smaller than the probability that a point will be made and we move to the second round of the game (which is 2/3).

To complete the tree, we need the probability that the game stops conditionally on each of the six different points that can appear in the come-out roll. Consider first the probability of winning if the point is 4. We can break this event down into winning in the first point roll after 4 becomes the point, or the second point roll, or the third point roll, and so on. All of these events are disjoint, therefore

$$P(\text{win} \mid \text{point is } 4) = P(\text{win in the first roll} \mid \text{point is } 4)$$

$$+ P\left(\text{win in the second roll} \mid \begin{array}{c} \text{did not win in} \\ \text{the first roll } \underline{\text{and}} \\ \text{point is } 4 \end{array}\right)$$

$$+ P\left(\text{win in the third roll} \mid \begin{array}{c} \text{did not win in the} \\ \text{first or second rolls} \\ \underline{\text{and}} \text{ point is } 4 \end{array}\right) \cdots$$

Now, the probability of winning in the first roll if the point is a 4 is simply the probability of getting a 4 in a roll of the dice

$$P(\text{win the first roll} \mid \text{point is } 4) = \frac{1}{12}.$$

On the other hand, to win in the second roll if you did not win in the first and the point is four, your first roll must have been anything but a 4 or a 7 and your second roll should be a 4. The probability that the first roll is not a 4 or a 7 is

$\frac{36-3-6}{36} = \frac{3}{4}$, while the probability that the second roll is 4 is 1/12. Since the rolls are independent, this means that

$$P\left(\text{win in the second roll}\,\middle|\,\begin{array}{l}\text{did not win in the}\\\text{first roll \underline{and} point is 4}\end{array}\right) = \frac{3}{4} \times \frac{1}{12}.$$

A similar argument can be used for subsequent rolls. In general, the probability of winning in k rolls if you have not won in the previous k and the point is 4 requires that you observe a series of outcomes that looks like

$$\underbrace{\text{X X X} \cdots \text{X}}_{k-1 \text{ times}} 4,$$

where X corresponds to any outcome of the dice that is not a 7 or a 4. This sequence has probability

$$P(\underbrace{\text{X X X} \cdots \text{X}}_{k-1 \text{ times}}4) = \left(\frac{3}{4}\right)^{k-1} \times \frac{1}{12}.$$

which leads to

$$P(\text{Win} \mid \text{point is 4}) = \underbrace{\frac{1}{12}}_{\substack{\text{Probability}\\\text{of winning in}\\\text{the 1st roll}}} + \underbrace{\frac{1}{12} \times \frac{3}{4}}_{\substack{\text{probability}\\\text{of winning in}\\\text{the 2nd roll}}} + \underbrace{\frac{1}{12} \times \left(\frac{3}{4}\right)^2}_{\substack{\text{probability}\\\text{of winning in}\\\text{the 3rd roll}}}$$

$$+ \underbrace{\frac{1}{12} \times \left(\frac{3}{4}\right)^3}_{\substack{\text{probability}\\\text{of winning in}\\\text{the 4}^{\text{th}}\text{ roll}}} + \cdots$$

$$= \frac{1}{12}\left\{1 + \frac{3}{4} + \left(\frac{3}{4}\right)^2 + \left(\frac{3}{4}\right)^3 + \left(\frac{3}{4}\right)^4 + \cdots\right\}.$$

Sums as the one in brackets are called *geometric sums* and appear very often when analyzing games that consist of sequences of independent trials. The following result is useful when dealing with geometric sums:

Finite Geometric Series

The sum of a geometric series with $n + 1$ terms is given by

$$1 + a + a^2 + \cdots + a^n = \frac{1 - a^{n+1}}{1 - a}$$

Note that, if $|a| < 1$ and n is very large then $a^{n+1} \approx 0$, which leads to

---

Infinite Geometric Series

The sum of an infinite geometric series is given by

$$1 + a + a^2 + a^3 + a^4 + \cdots = \frac{1}{1-a}$$

Accordingly,

$$1 + \frac{3}{4} + \left(\frac{3}{4}\right)^2 + \left(\frac{3}{4}\right)^3 + \left(\frac{3}{4}\right)^4 + \cdots = \frac{1}{1 - \frac{3}{4}} = 4,$$

and therefore

$$P(\text{win} \mid \text{point is 4}) = \frac{1}{12} \times 4 = \frac{1}{3}.$$

The following R code can be used to verify the formula for the infinite geometric sum:

```
> a   = 3/4
> n   = 20
> expon = seq(0,n)
> geometricseries = a^expon
> geometricseries   # Terms of the geometric series

 [1]  1.000000000 0.750000000 0.562500000 0.421875000
 [5]  0.316406250 0.237304688 0.177978516 0.133483887
 [9]  0.100112915 0.075084686 0.056313515 0.042235136
[13]  0.031676352 0.023757264 0.017817948 0.013363461
[17]  0.010022596 0.007516947 0.005637710 0.004228283
[21]  0.003171212

> cumsum(geometricseries) # Sum steadily approaches 4

 [1]  1.000000 1.750000 2.312500 2.734375 3.050781 3.288086
 [7]  3.466064 3.599548 3.699661 3.774746 3.831059 3.873295
[13]  3.904971 3.928728 3.946546 3.959910 3.969932 3.977449
[19]  3.983087 3.987315 3.990486

> (1 - a^(n+1))/(1-a)   # Same as last term of cumsum

[1]  3.990486
```

Another way to find the probability of winning when the point is 4 is to realize that, among the nine outcomes that end the game (three that add to 4 and six that add to 7), only the ones that add up to 4 make you win. This leads to $P(\text{win} \mid$

point is 4) = 3/9 = 1/3, just as before. The following simulation can be used to corroborate the calculations we just made:

```
> n = 10000
> result = rep(0,n)
> outspc = seq(1,6)
> point = 4
> for(i in 1:n){
+    dice = sample(outspc,2,replace=TRUE)
+    roll = sum(dice)
+    while(roll!=point & roll!=7){
+      dice = sample(outspc,2,replace=TRUE)
+      roll = sum(dice)
+    }
+    if(roll==point){
+      result[i] = "W"
+    }else{
+      result[i] = "L"
+    }
+ }
> sum(result=="W")/n

[1] 0.3357
```

A similar argument can be used for all the other points (and a small modification of the code above can be used to check them, see Exercise 14):

$$P(\text{win} \mid \text{point is 4}) = P(\text{win} \mid \text{point is 10}) = \frac{1}{3},$$

$$P(\text{win} \mid \text{point is 5}) = P(\text{win} \mid \text{point is 9}) = \frac{2}{5},$$

$$P(\text{win} \mid \text{point is 6}) = P(\text{win} \mid \text{point is 8}) = \frac{5}{11}.$$

The fully filled tree is presented in Figure 6.4. Now, following the Total Probability Law we discussed in Chapter 5, we can sum the probabilities associated

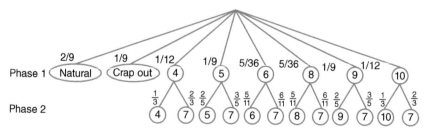

Figure 6.4 Tree representation for the possible results of the game of craps with the probabilities for all scenarios.

with the paths that lead you to win the come-out roll to get

$$P(\text{winning the pass line bet}) = P(\text{winning in the 1st phase})$$

$$+ P \left(\begin{array}{c} \text{winning in the 2nd} \\ \text{phase } \underline{\text{and}} \text{ point is 4} \end{array} \right) + P \left(\begin{array}{c} \text{winning in the 2nd} \\ \text{phase } \underline{\text{and}} \text{ point is 5} \end{array} \right)$$

$$+ P \left(\begin{array}{c} \text{winning in the 2nd} \\ \text{phase } \underline{\text{and}} \text{ point is 6} \end{array} \right) + P \left(\begin{array}{c} \text{winning in the 2nd} \\ \text{phase } \underline{\text{and}} \text{ point is 8} \end{array} \right)$$

$$+ P \left(\begin{array}{c} \text{winning in the 2nd} \\ \text{phase } \underline{\text{and}} \text{ point is 9} \end{array} \right) + P \left(\begin{array}{c} \text{winning in the 2nd} \\ \text{phase } \underline{\text{and}} \text{ point is 10} \end{array} \right)$$

Hence,

$$P(\text{winning the pass line bet}) = \underbrace{\frac{2}{9}}_{\substack{\text{Probability} \\ \text{winning in} \\ \text{1st phase}}} + \underbrace{\frac{1}{12}}_{\substack{\text{Probability} \\ \text{the point is} \\ \text{established} \\ \text{as 4}}} \times \underbrace{\frac{1}{3}}_{\substack{\text{Probability} \\ \text{winning in 2nd} \\ \text{phase given} \\ \text{the point is 4}}}$$

$$+ \underbrace{\frac{1}{9}}_{\substack{\text{Probability} \\ \text{the point is} \\ \text{established} \\ \text{as 5}}} \times \underbrace{\frac{2}{5}}_{\substack{\text{Probability} \\ \text{winning in 2nd} \\ \text{phase given} \\ \text{the point is 5}}} + \underbrace{\frac{5}{36}}_{\substack{\text{Probability} \\ \text{the point is} \\ \text{established} \\ \text{as 6}}} \times \underbrace{\frac{5}{11}}_{\substack{\text{Probability} \\ \text{winning in 2nd} \\ \text{phase given} \\ \text{the point is 6}}}$$

$$+ \underbrace{\frac{5}{36}}_{\substack{\text{Probability} \\ \text{the point is} \\ \text{established} \\ \text{as 8}}} \times \underbrace{\frac{5}{11}}_{\substack{\text{Probability} \\ \text{winning in 2nd} \\ \text{phase given} \\ \text{the point is 8}}} + \underbrace{\frac{1}{9}}_{\substack{\text{Probability} \\ \text{the point is} \\ \text{established} \\ \text{as 9}}} \times \underbrace{\frac{2}{5}}_{\substack{\text{Probability} \\ \text{winning in 2nd} \\ \text{phase given} \\ \text{the point is 9}}}$$

$$+ \underbrace{\frac{1}{12}}_{\substack{\text{Probability} \\ \text{the point is} \\ \text{established} \\ \text{as 10}}} \times \underbrace{\frac{1}{3}}_{\substack{\text{Probability} \\ \text{winning in 2nd} \\ \text{phase given} \\ \text{the point is 10}}} ,$$

which simplifies to $P(\text{winning the pass line bet}) = 244/495 \approx 0.4929$. Furthermore, since the pass line bet pays even odds, the expected profit for every dollar invested is

$$E(\text{profit from pass line bet}) = (-1) \times \frac{251}{495} + 1 \times \frac{244}{495} \approx -0.01414.$$

As you can see, the house advantage for the pass line bet in craps is much smaller than the house advantage in roulette! On purely monetary terms then, there is no reason for you to ever play roulette again!

We can extend the code we used before to check the probability of winning given that the point is 4 to corroborate the probability of winning when playing the pass line bet.

```
> n = 100000
> result = rep(0,n)
> outspc = seq(1,6)
> for(i in 1:n){
+     dice = sample(outspc,2,replace=TRUE)   #First round
+     roll = sum(dice)
+     if(roll %in% c(4,5,6,8,9,10)){   #Second round
+       point = roll
+       dice = sample(outspc,2,replace=TRUE)
+       roll = sum(dice)
+       while(roll!=point & roll!=7){
+          dice = sample(outspc,2,replace=TRUE)
+          roll = sum(dice)
+       }
+       if(roll==point){
+          result[i] = "W"   #Win in second round
+       }else{
+          result[i] = "L"   #Lose in second round
+       }
+     }else{
+       if(roll==7 | roll==11){
+          result[i] = "W"   #Win in first round
+       }else{
+          result[i] = "L"   #Lose in first round
+       }
+     }
+ }
> sum(result=="W")/n

[1] 0.49454

> mean( (result=="W") - (result=="L") )

[1] -0.01092
```

6.1.2 The Don't Pass Line Bet

The *don't pass line* (also called *lose* or *wrong*) bet is a wager against the shooter that is almost the mirror image of the pass line bet and also pays even odds. The don't pass line bet is always played as a stand-by bet that runs parallel to the pass line bet and is resolved in the following way. If the come-out roll made by the shooter is a 7 or 11, then the don't pass line bet loses automatically. On the other hand, if the come-out roll is a 2 or 3, then the don't pass line bet wins automatically, but if the come-out is a 12 then the game ends in a tie (this is sometimes called a *push*) and the player gets her original bet back. Finally, if a point is made, the player betting on the don't pass line wins if a 7 comes up first and loses if the point comes up.

Note that, except for the outcome for 12 in the come-out roll, the don't pass line bet is the opposite to the pass line bet. Hence, we can easily compute the probability of winning and the probability of tying this bet using a similar procedure to the one outlined in the previous section. This leads to

$$P(\text{winning the don't pass line bet}) = \frac{949}{1980} \approx 0.47929,$$

$$P(\text{push in the don't pass line bet}) = \frac{1}{36} \approx 0.02778,$$

$$P(\text{losing in the don't pass line bet}) = \frac{244}{495} \approx 0.49293,$$

and therefore

$$E(\text{profit from don't pass line bet}) = (-1) \times \frac{244}{495} + 0 \times \frac{1}{36} + 1 \times \frac{949}{1980}$$
$$\approx -0.01364.$$

Note that the don't pass line bet is slightly less disadvantageous to the player than the pass line bet! You can modify the simulation of the pass line bet we provided in the previous section to check these results (see Exercise 16).

6.1.3   The Come and Don't Come Bets

In addition to the pass and don't pass line bets, there are two more line bets usually offered by casinos: the *come* and the *don't come* bets. A come bet works almost identically to the pass line bet, but it is played out of synchrony with and independently of it. As soon as the player makes a come bet, it starts its own first phase regardless whether the shooter is playing his come-out roll or a point roll. Consequently, if a 7 or 11 is rolled on the first round after the player placed the chips in the come area of the table, it wins, but if a 2, 3, or 12 is rolled, it loses. On the other hand, if the roll is 4, 5, 6, 8, 9, 10 then the come bet will be moved by the base dealer onto a Box representing the number the shooter threw. The don't come bet is similar, but it mirrors the don't pass line bet instead.

6.1.4   Side Bets

In addition to line and come bets, many casinos allow for *single-roll* and *multi-roll* bets. These bets can be placed at any time by either the shooter or any stand-by player. Examples of a single-roll bet include *snake eyes* (which involves betting on two ones coming up, typically paying 30 to 1) and the *Yo* (which involves betting on 11 coming up and often pays 15 to 1). The analysis of these bets is very similar to the analysis of the bets in roulette. For example, the expected profit from a snake eyes bets is

$$E(\text{profit from snake eyes}) = 30 \times \frac{1}{36} + (-1) \times \frac{35}{36} = -\frac{5}{36} = -0.13889,$$

while the expected profit for Yo is

$$E(\text{profit from Yo}) = 15 \times \frac{2}{36} + (-1) \times \frac{34}{36} = -\frac{4}{36} = -0.11111.$$

Note that the house advantage for these two bets is much larger than for the line bets available in craps (as well as for any of the bets available in roulette). Hence, these side bets are usually very bad proposition for the player.

An example of *multi-roll* bets is the *hard way* bet, in which the player bets that the shooter will throw a 4, 6, 8, or 10 the hard way (recall Table 6.1) before he throws a 7 or the corresponding easy way. We can use some of the ideas discussed in this chapter to compute house advantage for these bets. For example, in the case of the *hard 8* bet (which pays 9 to 1 on a winning bet), note that there is only one dice combination that makes you win (two 4s), while there are 10 combinations that make you lose (6 ways in which 7 can happen, plus 3 and 5, 5 and 3, 2 and 6, and 6 and 2). Since any other number just forces you to continue rolling, the probability of winning this bet is 1/11, the probability of losing is 10/11, and the expected profit is

$$E(\text{profit from hard 8}) = 9 \times \frac{1}{11} + (-1) \times \frac{10}{11} = -\frac{1}{11} = -0.090909.$$

Again, this bet is quite bad for the players!

6.2 Exercises

1. When playing craps, what is the probability of crapping out?

2. When playing craps, if your point is 9, what is the probability that you will win within the next four shots? What is the probability that you will lose within the next four shots?

3. When playing craps, if your point is 5, what is the probability that you will win within the next four shots? What is the probability that you will lose within the next four shots?

4. If your point is 6, what is the probability that you will win the round of craps? Show your calculations.

5. If your point is 10, what is the probability that you will win the round of craps? Show your calculations.

6. If your point is 11, what is the probability that you will win the round of craps? Show your calculations.

7. Show that the probability of winning the don't pass line bet is 949/1980.

8. Verify the value provided in the text for the house advantage in the don't pass line bet.

9. The don't pass line bet in craps is meant to be a bet against the shooter. Indeed, the don't pass line bet loses if the shooter gets a 7 or 11 in the come-out roll, or if the shooter gets the point during the follow-up rolls. However, the don't pass line bet wins in the come-out roll only if the shooter gets 2 or 3 but ties if the shooter gets a 12. Why are the rules setup in this way instead of just letting the don't pass line bets win for all come-out rolls that are 2, 3, or 12?

10. Why does a casino allow players to retire (take back) a don't pass line bet after the first roll has been made, yet it does not let you do the same for pass line bets?

11. What is the probability of winning the come and don't come bets?

12. In craps, the *field* bet is a single-roll wager in which the player wins if the next roll is 2, 3, 4, 9, 10, 11, or 12, and losses on any other number. The typical payout for this bet is 1 to 1 if 3, 4, 9, 10 and 11, 2 to 1 on a 2, and 3 to 1 on a 12. What is the house advantage for this bet, and how does it compare with the house advantage in the pass line bet?

13. What is the house edge on a *hard 6* bet? How does it compare against the house edge for a hard 8 bet?

14. [R] Find out the value of the cumulative sums associated with the series $1 + \frac{1}{3} + \frac{1}{3}^2 + \frac{1}{3}^3 + \frac{1}{3}^4 + \cdots$ using R.

15. [R] Create a simulation to compute the probability of winning the pass-line bet if the point is 9.

16. [R] Create a simulation to compute the probability of winning the don't pass line bet.

7

Roulette Revisited

In this chapter, we make use of some of the concepts we learned so far to address additional interesting questions about roulette. Although we concentrate on roulette, many of the ideas discussed here can be extended to other games based on independent rounds such as craps.

7.1 Gambling Systems

Gambling systems are strategies that increase or decrease the size of a bet according to whether the player is winning or losing. They are promoted as tools that allow players to beat the house advantage; plenty of books have been written on the topic; and many gamblers have rediscovered the same tactics over and over again. However, it is important to emphasize that for games that rely on independent rounds of play such as roulette or craps, *no system can be devised to beat the house.*

7.1.1 Martingale Doubling Systems

Martingale doubling systems are very simple. To play the system, you must keep betting until you win, doubling your bet every time you lose. More specifically, you start a betting cycle by making a small bet, such as $1. If you lose, you double your bet and gamble again; if you win, you take your winnings and start a new cycle by betting $1. Typically even bets, such as the color bet in roulette, are used. However, this is not a requirement.

In a world where you can keep making bets indefinitely, the martingale doubling system guaranties that you will make $1 at the end of a betting cycle no matter what the probability of winning is. Indeed, say that it takes n individual bets to complete a cycle. Since the bet is even, at that point you win $2. Also, since you lost the previous $n - 1$ wagers, you have lost a total of

Probability, Decisions and Games: A Gentle Introduction using R, First Edition. Abel Rodríguez and Bruno Mendes.
© 2018 John Wiley & Sons, Inc. Published 2018 by John Wiley & Sons, Inc.
Companion website: www.wiley.com/go/Rodriguez/Probability_Decisions_and_Games

$1 + 2 + 2^2 + \cdots + 2^{n-1}$. Note that the amount you have lost is the sum of terms of a geometric series. Recall from Chapter 6 that

$$1 + a + a^2 + \cdots + a^{n-1} = \frac{1 - a^n}{1 - a}$$

In this case, we have $a = 2$. Therefore,

$$1 + 2 + 2^2 + \cdots + 2^{n-1} = \frac{1 - 2^n}{1 - 2} = 2^n - 1.$$

and your profit for the cycle is

$$\text{Profit} = \underbrace{2^n}_{\substack{\text{Winnings from the last} \\ \text{(winning) bet}}} - \underbrace{(2^n - 1)}_{\substack{\text{Accumulated losses from} \\ \text{previous (losing) bets}}} = 1$$

no matter what the value of n is.

At the first sight, this calculation suggests that a doubling system should allow you to always make money. What can go wrong? The underlying assumption of this system is that you can keep playing indefinitely until you win. However, in real life, your bankroll is finite, and the bets you need to make to keep the system going grow very fast. So you might not be able to cover the next bet required by the system to keep going, at which point you will lose all your money.

To illustrate this, assume that you have $1000 and your initial bet is $1. How many losses in a row can you take before you run out of money to make the next bet? We just showed that the accumulated loss after n rounds of the game is $2^n - 1$ (see also Table 7.1). Hence, if we lose 9 times in a row, we will only have $489 left, which is not enough money to cover the 10th bet that the system

Table 7.1 Accumulated losses from playing a martingale doubling system with an initial bet of $1 and an initial bankroll of $1000.

Round (n)	Bet on this round	Accumulated loss	Money left
1	1	0	1000
2	2	1	999
3	4	3	997
4	8	7	993
5	16	15	985
6	32	31	969
7	64	63	937
8	128	127	873
9	256	255	745
10	512	511	489

requires (which would be for $512)! In general, the number of rounds that you can play is simply given by $\lfloor \log_2 B \rfloor$ where B is the amount of money in your bankroll and $\lfloor x \rfloor$ means "round the number x down to the nearest integer" (in this case, $\log_2 1000 \approx 9.9658$, so $\lfloor \log_2 1000 \rfloor = 9$). This expression also makes it clear that doubling your bankroll (e.g., taking your initial money from $1000 to $2000) only buys you one additional round before you go bust!!

Now, you may argue that losing 9 times in a row when making color bets in roulette is a very unlikely event. Because the spins of the roulette wheel are independent, the exact probability of this happening is

P(losing 1st spin and losing 2nd spin $\cdots$ and losing 9th spin)

$$= P(\text{losing 1st spin}) \times P(\text{losing 2nd spin}) \times \cdots \times P(\text{losing 9th spin})$$

and therefore

$$P(\text{losing 9 times in a row}) = \underbrace{\frac{20}{38} \times \frac{20}{38} \times \cdots \times \frac{20}{38}}_{9 \text{ times}} \approx 0.003098972.$$

Consequently, even if we start with $1000 and bet only $1 initially, we have that roughly every 300 cycles we will not be able to cover the next bet and the martingale system will fail (the exact number is $1/0.003098972 = 322.6877$ cycles). In the meantime, we would have made a profit of about $300, but even if we reinvest the winnings we are bound to eventually run out of money to cover the next bet required by the system.

Some additional intuition can be obtained by simulating the running profit of playing a martingale doubling system with an initial bet $1 over 2000 spins of the roulette:

```
> spins = 2000
> outspc = c("W","L")
> outpro = c(18/38, 20/38)
> profit = rep(0,spins)
> bet = 1
> for(i in 1:spins){
+    outcome = sample(outspc,1,replace=TRUE,prob=outpro)
+    if(outcome=="W"){
+      profit[i] = bet
+      bet = 1
+    }else{
+      profit[i] = -bet
+      bet  = 2*bet
+    }
+ }
> plot(cumsum(profit), type="l", xlab="Spin",
+                       ylab="Cumulative Profit")
> abline(h=0, lty=2)
```

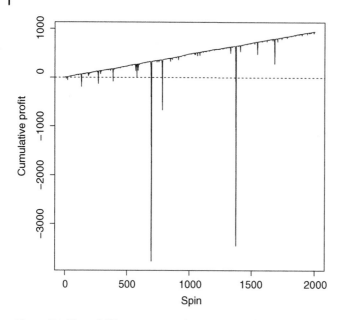

Figure 7.1 The solid line represents the running profits from a martingale doubling system with $1 initial wagers for an even bet in roulette. The dashed horizontal line indicates the zero-profit level.

Figure 7.1 shows the result of one such simulation. The increasing trend in the cumulative profit suggests that, as advertised, the system makes money as long as we can keep playing it indefinitely. Note, however, that the positive trend is punctuated by sporadic big losses (of over $2000 in one case, even though our initial bet was only $1). It is these big sporadic loses that make the system fail in real life!

7.1.2 The Labouchère System

To play the *Labouchère system*, you need to decide how much money you want to win and then write a list of positive numbers that add up to that quantity. For the sake of argument, say that you want to make $100, and you decide to use the numbers 15, 15, 20, 25, 20, 5 in your game. You always bet the sum of the first and last numbers in the list (if a single number remains, you use that number). If you win, the two numbers are removed from the list; if you lose, the amount of the losing bet is added at the end of the list. You stop playing once there are no more numbers in the list.

First, you need to convince yourself that the system, if completed, will indeed allow you to win the sum of the amounts in the list. To see this, assume first that it is your lucky day and you win all your bets straight. In our example, that means that the first time you bet (and win) $15 + 5 = \$20$, the second time you win $15 + 20 = \$35$, and the third time you win $20 + 25 = \$45$. So the total amount you win is $20 + 35 + 45 = \$100$ as expected. What if you lose your first bet but then win all others straight? In that case, your list now contains the numbers 15, 15, 20, 25, 20, 5, 20 and you are losing \$20. But if you now win all your bets in a row you will be making \$120, so your net profit will be \$100 again. In general, by adding the amounts you lose to the end of the list you make up for any loses you might have incurred in the middle of the game before stopping, which ensures that you will make the desired amount of money. The following R code simulates the running profit of the Labouchère based on an initial list with 50 entries of \$10 each used on an even roulette bet. The cumulative profit from the system can be seen in Figure 7.2.

```
> outspc = c("W","L")
> outpro = c(18/38, 20/38)
> listlength = 50
> betvalue   = 10
> listofbets = rep(betvalue, listlength)
> profit = 0
> while(length(listofbets)>0){
+    if(listlength==1){
+      currentbet = listofbets[1]
+    }else{
+      currentbet = listofbets[1] + listofbets[listlength]
+    }
+    outcome = sample(outspc,1,replace=TRUE,prob=outpro)
+    if(outcome=="W"){
+      profit = c(profit, currentbet)
+      listofbets = listofbets[-c(1,listlength)]
+    }else{
+      profit = c(profit, -currentbet)
+      listofbets = c(listofbets, currentbet)
+    }
+    listlength = length(listofbets)
+ }
> plot(cumsum(profit), type="l", xlab="Spin",
+                        ylab="Cumulative Profit")
> abline(h=0, lty=2)
```

Just like the martingale doubling system, the Labouchère system would seem to ensure that you always make money when playing roulette. However, Labouchère systems share the same weaknesses as martingale doubling systems. If you hit a bad enough losing streak you might run out of money before you have the chance to recoup your previous loses or make any money. However, as the simulation suggests, since the size of the bets in a Labouchère system grow linearly rather than exponentially, the number of games you are able to play before going bankrupt tends to be larger.

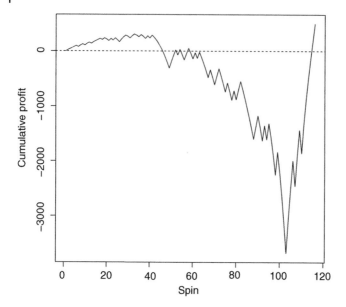

Figure 7.2 Running profits from a Labouchère system with an initial list of $50 entries of $10 for an even bet in roulette. Note that the simulation stops when the cumulative profit is 50 × 10 = 500; the number of spins necessary to reach this number will vary from simulation to simulation.

7.1.3 D'Alembert Systems

The *D'Alembert system* is based on the idea that a win is less likely if you have just won and more likely if you have just lost. Hence, you should increase the amount of your bet after you lose and reduce it after you win. The recommended progression is typically linear, so that you add a fixed quantity (say, $1) to your bet when you lose, and subtract the same quantity every time you win, to the table minimum.

Whereas the martingale doubling systems and the Labouchère systems are based on mathematically sound principles (they do not work only because in real life we do not have an infinite amount of money in the bank), the D'Alembert system is based on an erroneous probabilistic argument. The spins of the roulette are independent from each other, which means that the probability of winning or losing does not depend on the past (the game is *memoryless*). It is true that, before you make any spin, the probability of getting 9 loses in a row when playing even bets in roulette is very small. However, after you have already seen 8 loses, the probability of getting the 9th is exactly the same as the probability of getting the first one.

The following R code simulates the cumulative profit from applying the D'Alembert system with an initial bet of $5, change in bets of $1, minimum bet of $1, and maximum bet of $20 to an even roulette bet. The clear decreasing trend and large negative values in Figure 7.3 corroborate our argument that the D'Alembert system does not work.

```
> n = 10000
> outspc = c("W","L")
> outpro = c(18/38, 20/38)
> profit = rep(0,n)
> currentbet = 5
> incrementbet = 1
> minimumbet = 1
> maximumbet = 20
> for(i in 1:n){
+    outcome = sample(outspc,1,replace=TRUE,prob=outpro)
+    if(outcome=="W"){
+       profit[i] = currentbet
+       currentbet = max(currentbet-1, minimumbet)
+    }else{
+       profit[i] = -currentbet
+       currentbet = min(currentbet+1, maximumbet)
+    }
+ }
> plot(cumsum(profit), type="l", xlab="Spin",
+                          ylab="Cumulative Profit")
> abline(h=0, lty=2)
```

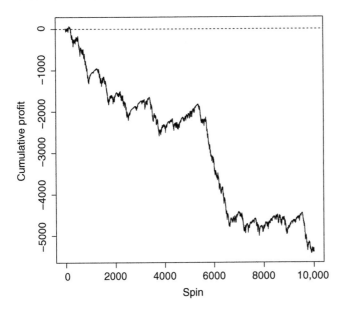

Figure 7.3 Running profits over 10,000 spins from a D'Alembert system with an initial bet of $5, change in bets of $1, minimum bet of $1 and maximum bet of $20 to an even roulette bet.

7.2   You are a Big Winner!

Even though the expected profit in roulette is negative, it is actually not uncommon for players to be able to get ahead for a while. Indeed, you can temporarily make a lot of money in roulette, but the law of large numbers implies that if you want to keep it, you need to stop playing and never do it again for the rest of your life!

For example, let's compute the probability of winning exactly 10 rounds out of 15 played when making $1 color bets (that would mean that you are ahead by $5 after playing 15 rounds). There are many ways in which this could happen; for example, you could win the first 10 rounds and lose the next 5,

W W W W W W W W W W L L L L L,

or you could lose the 2nd, 3rd, 5th, 12th, and 13th,

W L L W L W W W W W W L L W W.

Consider first the probability of each one of these sequences. Since the rounds are independent, all sequences of 15 spins that include 10 wins and 5 loses have the same probability,

(recall that the probability of winning a color bet is $\frac{18}{38}$, while the probability of losing it is $\frac{20}{38}$). This means that

$$P(\text{W W W W W W W W W W L L L L L})$$

$$= P(\text{W L L W L W W W W W W L L W W}) = \left(\frac{18}{38}\right)^{10} \times \left(\frac{20}{38}\right)^{5}.$$

Now, to compute the total probability of winning 10 rounds out of 15, we need to sum the probabilities of all sequences that match the criteria. Since all of the different sequences have the same probability, this boils down to counting the number of sequences that match the criteria.

To compute the total number of ways in which you can get 10 wins in 15 spins, recall again the combinatorial numbers we discussed in Chapter 4. We need to pick 10 positions in the list out of 15, and the order in which the 10 positions are selected is of no consequence to us. Therefore, there are $\binom{15}{10} = \frac{15!}{10! \times 5!} = 3003$ ways in which you can have 10 wins in 15 spins of a roulette. Hence,

$$P(\text{winning 10 rounds out of 15}) = \binom{15}{10} \times \left(\frac{18}{38}\right)^{10} \times \left(\frac{20}{38}\right)^{5}.$$

More generally, consider a random variable Z that counts the number of wins out of n rounds. The same argument we used before leads to

$$P(Z = k) = \binom{n}{k} \times \left(\frac{18}{38}\right)^{k} \times \left(\frac{20}{38}\right)^{n-k}.$$

Note that if $Z = k$, then you made \$$k$ from the rounds you won, and lost \$$(n-k)$ from those you lost, making your profit from playing the game $k - (n - k) = 2k - n)$ dollars.

Now let's put this result to good use. Say that you have been playing roulette all night. For the sake of the argument, say you have played 300 rounds (which means about 5 hours at a rate of 60 spins an hour) by betting \$1 each time on color. After all this, you are ahead by \$20 (this means you had 20 more wins than losses) and you feel very unlucky because you have made so little money. Are you justified?

One way to address this question is to compute the probability that somebody would win \$20 or more after 300 games. Now, for you to be ahead by \$20 or more, you would need to win at least 160 of the 300 rounds you have played, so we need to compute

$$\underbrace{P(Z \geq 160)}_{\substack{\text{Probability of winning} \\ \text{160 or more} \\ \text{roulette spins}}} = \underbrace{\binom{300}{160} \times \left(\frac{18}{38}\right)^{160} \times \left(\frac{20}{38}\right)^{140}}_{\substack{\text{Probability of winning} \\ \text{160 rounds out of 300}}} +$$

$$\underbrace{\binom{300}{161} \times \left(\frac{18}{38}\right)^{161} \times \left(\frac{20}{38}\right)^{139} + \cdots +}_{\substack{\text{Probability of winning} \\ \text{161 rounds out of 300}}} \underbrace{\binom{300}{300} \times \left(\frac{18}{38}\right)^{300} \times \left(\frac{20}{38}\right)^{0}}_{\substack{\text{Probability of winning 300} \\ \text{rounds out of 300}}}.$$

Computing this quantity by hand is difficult, but you can use R to obtain the number (see Sidebar 7.1):

```
> pbinom(159, size=300, prob=18/38, lower.tail=FALSE)

[1] 0.022217
```

This means that, for every 100 players, only about 2 would have made \$20 or more after playing for 5 hours... I would consider you quite lucky!

7.3 How Long will My Money Last?

We can use some of the tools we developed to study the martingale doubling system to answer other interesting questions about roulette. For example, suppose that you want to go out and play roulette tonight. Since the expected profit from this game is negative, you know for sure that you will eventually lose all your money. However, how long you play is a random variable whose distribution will depend on how much money you have and how much you bet each time.

To make things simple for now, say that you start with just $1, that you bet $1 each round, that you do not reinvest your winnings, and that you try to make your original $1 last for as long as possible by playing even bets such as a *color* bet. If you are a very unlucky individual, you might lose on the first spin, so that you might be able to play only one round. So, if you let

$$X = \{\text{Number of rounds you play if you have \$1 and bet it all}\},$$

then we have $P(X = 1) = 20/38$.

Now, for you to be able to play exactly two rounds, you would need to win the first round and lose the second. Therefore, since spins are independent, we have $P(X = 2) = \frac{18}{38} \times \frac{20}{38}$. More generally, for you to play exactly k rounds, you need to win the first $k - 1$ rounds and lose in the kth round, which happens with probability

$$P(X = k) = \left(\frac{18}{38}\right)^{k-1} \times \frac{20}{38},$$

where k could be any integer number greater or equal than 1. Table 7.2 shows a graph of probability as a function of k; as you would expect, the longer the streak, the lower its probability.

To compute the average length of one such streak, that is, $E(X)$, you would need to compute

$$E(X) = \frac{20}{38} + 2 \times \frac{18}{38} \times \frac{20}{38} + 3 \times \left(\frac{18}{38}\right)^2 \times \frac{20}{38} + 4 \times \left(\frac{18}{38}\right)^3 \times \frac{20}{38} + \cdots$$

Table 7.2 Probability that you play exactly *k* rounds before you lose your first dollar for *k* between 1 and 6.

k	P(X = k)
1	0.5263158
2	0.2493075
3	0.1180930
4	0.0559388
5	0.0264973
6	0.0125514

which can be rewritten as

$$E(X) = \frac{20}{38} + \frac{18}{38} \times \frac{20}{38} + \left(\frac{18}{38}\right)^2 \times \frac{20}{38} + \left(\frac{18}{38}\right)^3 \times \frac{20}{38} + \cdots$$
$$+ \frac{18}{38} \times \frac{20}{38} + \left(\frac{18}{38}\right)^2 \times \frac{20}{38} + \left(\frac{18}{38}\right)^3 \times \frac{20}{38} + \cdots$$
$$+ \left(\frac{18}{38}\right)^2 \times \frac{20}{38} + \left(\frac{18}{38}\right)^3 \times \frac{20}{38} + \cdots$$
$$+ \left(\frac{18}{38}\right)^3 \times \frac{20}{38} + \cdots$$
$$\vdots$$

With a little bit of algebra, and using again the formula for the sum of the terms of a geometric series on each row, we get $E(X) = \frac{38}{20} \approx 1.9$. Accordingly, on an average night, you would play for a little bit less than two rounds!

The previous scenario is probably too simple to be of practical use. For example, even if you decide not to reinvest your winnings, you would probably not go to the table with only $1. So, let's say that you start with $10, and you make $1 bets (but do not reinvest your winnings). There are a couple of ways in which you can work with the random variable

$Y = \{$Number of rounds you play if you have $10 and make $1 bets$\}$.

If you only care about the expectation, you can proceed in the following way. Since you make $1 bets, you can think about the gambling process as making 10 bets of $1, and riding each one until you lose the dollar. This means that you can write

$$Y = X_1 + X_2 + X_3 + \cdots + X_9 + X_{10},$$

where each X_i corresponds to one independent realization of our original random variable. Therefore, we can easily see that

$$E(Y) = E[X_1] + E[X_2] + E[X_3] + \cdots + E[X_9] + E[X_{10}] = 19.$$

In other words, if you start with $10 and make $1 bets, you can expect to play for about 38 minutes on an average night (assuming about one spin every 2 minutes).

If you care about the whole distribution of Y, the following approach is a bit simpler than dealing with the sum of multiple random variables. For you to play exactly k rounds before losing all your money, you need a sequence of wins and losses that satisfies two conditions: (1) there are exactly 10 losses, and (2) the kth round (the last one) is a loss. In other words, you need a sequence such as

L W W W W W L L L W W L W L W W W L L W L W W W L.

Sidebar 7.1 The Binomial Distributions in R

R includes functions that allow you to compute probabilities associated with a host of well known random variables. For example, the functions `dbinom()` and `pbinom()` allow you to compute $P(Z = x)$ and either $P(Z \leq x)$ or $P(Z > x)$ when Z follows a binomial distribution. For example, for a binomial distribution with $n = 10$ and $p = 18/38$

$$P(Z = 4) = \binom{10}{4} \left(\frac{18}{38}\right)^4 \left(\frac{20}{38}\right)^6$$

which can be computed in either of these two forms:

```
> choose(10,4)*(18/38)^4*(20/38)^6
```

```
[1] 0.224726
```

```
> dbinom(4, size=10, prob=18/38)
```

```
[1] 0.224726
```

On the other hand, for $P(Z \leq 4) = \sum_{x=4}^{10} \binom{10}{x} \left(\frac{18}{38}\right)^x \left(\frac{20}{38}\right)^{10-x}$

```
> pbinom(4, size=10, prob=18/38)
```

```
[1] 0.4431709
```

while $P(Z > 4) = 1 - P(Z \leq 4)$ is obtained as

```
> pbinom(4, size=10, prob=18/38, lower.tail=FALSE)
```

```
[1] 0.5568291
```

Finally, the function `rbinom()` can be used to generate random numbers that follow the binomial distribution:

```
> rbinom(12, size=10, prob=18/38)
```

```
[1] 5 3 7 3 6 2 2 6 8 2 4 5
```

Now, this sequence is of length 25 and (since rounds are taken together and are independent from each other) it has probability

$$\left(\frac{18}{38}\right)^{15} \times \left(\frac{20}{38}\right)^{10}.$$

However, note that this is not the only possible sequence that satisfies these criteria. As a matter of fact, there are $\binom{24}{9} = \frac{24!}{9! \times 15!}$ such sequences (recall our discussion on combinatorial numbers from Chapter 4 and notice the last position has to be a loss, so we need to pick 9 positions for the remaining losses among 24 options). Since all of these sequences have the same probability:

$$P(Y = 25) = \binom{24}{9} \times \left(\frac{18}{38}\right)^{15} \times \left(\frac{20}{38}\right)^{10}.$$

More generally, the probability that you are able to play for k rounds if you started with $\$n$ and you bet $\$1$ per round and do not reinvest your winnings is

$$P(Y = k) = \binom{k - 1}{n} \times \left(\frac{18}{38}\right)^{k-n} \times \left(\frac{20}{38}\right)^{n}$$

for any $k \geq n$.

The case in which winnings are reinvested is a bit trickier and beyond the scope of this book. However, a simulation in R can provide you with some intuition:

```
> n = 10000
> outspc = c("W","L")
> outpro = c(18/38, 20/38)
> numspins = rep(0,n)
> for(i in 1:n){ # Simulation assumes profits get reinvested
+    bank = 10
+    spins = 0
+    while(bank>0){
+       spins = spins + 1
+       outcome = sample(outspc,1,replace=TRUE,prob=outpro)
+       bank = bank - (outcome=="L") + (outcome=="W")
+    }
+    numspins[i] = spins
+ }
> mean(numspins)    # Average number of spins

[1] 191.8288

> max(numspins)    # Maximum number of spins observed

[1] 4324
```

Note that by reinvesting your winnings you can significantly prolong the amount of time your $10 will last. However, since the expected value of the game is negative, you are bound to eventually go bankrupt!

7.4 Is This Wheel Biased?

In Chapter 3, we discussed the use of Chebyshev's inequality to approximately determine the number of spins needed to detect bias in a wheel. We consider now the related question of whether a sample consisting of a given number of spins provides evidence of a biased wheel. For example, let's assume that you have collected the results of 10,000 roulette spins and you observe that the number 31 has appeared 270 times (recall you would have expected to see it just about $10{,}000 \times \frac{1}{38} \approx 263$ times in this many spins). Does this suggest that the roulette is biased in favor of the number 31?

To answer this question, let's compute the probability that you observe the number 31 at least 270 times in 10,000 spins of the wheel *if the roulette is not biased*,

$$
P\begin{pmatrix} \text{Number 31 coming up 270} \\ \text{or more times in 10,000} \\ \text{spins of an unbiased wheel} \end{pmatrix} = \binom{10{,}000}{270} \times \left(\frac{1}{38}\right)^{270} \times \left(\frac{37}{38}\right)^{9730}
$$

$$
+ \binom{10{,}000}{271} \times \left(\frac{1}{38}\right)^{271} \times \left(\frac{37}{38}\right)^{9729} + \cdots +
$$

$$
\binom{10{,}000}{10{,}000} \times \left(\frac{1}{38}\right)^{10{,}000} \times \left(\frac{37}{38}\right)^{0}.
$$

This leads to

$$
P\begin{pmatrix} \text{Number 31 coming up 270 or} \\ \text{more times out of 10,000 spins} \end{pmatrix} \approx 0.3429.
$$

As before, you can use R to compute this number

```
> pbinom(269, size=10000, prob=1/38, lower.tail=FALSE)

[1] 0.3429242
```

Since this number is relatively large, there is little reason to think that the wheel is biased (the difference between 263 and 270 is small enough for it to be likely due to randomness).

7.5 Bernoulli Trials

When you look at the outcomes of multiple rounds of roulette, you are looking at an example of a very particular type of experiment called *Bernoulli trials*. A set of Bernoulli trials satisfies the following requirements:

- Each repetition of the experiment is independent from the rest.
- There are only two possible outcomes for each repetition of the experiment (call them win and lose).
- The probabilities of winning and losing are the same for each experiment (call the probability of success p).

There are a number of interesting probability distributions associated with Bernoulli trials. These distribution appeared in previous sections. For example, the *binomial distribution* arises when we are interested in the number of wins k among a total of n repetitions of the experiment.

The binomial distribution is given by

$$P(Z = k) = \binom{n}{k} \times p^k \times (1 - p)^{n-k}, \qquad\qquad k = 1, 2, 3, \ldots, n$$

For example, the binomial distribution appears when computing the probability that a certain outcome of roulette appeared k times in n repetition of the wheel (recall Section 7.4).

The *geometric distribution* arises when we want to know how many trials it will take us to get one success,

---

The geometric distribution is given by

$$P(X = n) = p \times (1 - p)^{n-1}, \qquad\qquad n = 1, 2, 3, \ldots$$

On the other hand, the *negative binomial distribution* appears when we want to know how many trials it will take us to get k successes (therefore, the geometric distribution is a special case of the negative binomial when $k = 1$).

The negative binomial distribution is given by

$$P(Y = n) = \binom{n-1}{k-1} \times p^k \times (1 - p)^{n-k}, \qquad\qquad n = k, k + 1, k + 2, \ldots$$

The geometric and negative binomial distributions appeared in Section 7.3 when we investigated the number of spins of a roulette wheel that you can make before running out of money if profits are not reinvested. Note that the main difference between the binomial and the negative binomial random variables is what is considered fixed and what is considered random. While the binomial distribution assumes that the number of trials n is fixed and the number of wins k is random, the negative binomial assumes the opposite.

7.6 Exercises

1. What is the martingale doubling system? How can it fail?

2. What is the Labouchère system? How can it fail?

3. How can minimum and maximum table bets affect the likelihood that you go bust when using a martingale doubling system?

4. Would a *martingale tripling* system avoid the problems with the martingale doubling system?

5. You decide to play roulette using the martingale doubling system. If your bankroll is $30, your initial bet is $1 and you do not reinvest your winnings, what is the average amount of time you might expect to play?

6. You decide to play roulette using a martingale tripling system. If your bankroll is $90, your initial bet is $1 and you do not reinvest your winnings, what is the average amount of time you might expect to play?

7. How could you use the martingale doubling system in craps?

8. What is the probability of winning 12 even bets in 30 spins of the roulette?

9. What is the probability of winning 12 even bets in 200 spins of the roulette?

10. The probability of getting a 7 in the game of craps is 1/6. The famous craps player known as the *dice dominator* is said to have avoided a 7 in the point-phase of the game of craps for 30-something consecutive rolls. To keep it simple, just imagine rolling two dice and you are only interested in whether a 7 comes out or not; what is the probability of avoiding 7 in 35 consecutive rolls of a die?

11. Do you think you can be called a *dice dominator* if you can avoid 7 for 15 consecutive rolls of two dice?

12. Say that you are trying to determine if a given (European) roulette wheel is biased in favor of the number 16. To do that you collect the outcome of 15,000 spins, and find that 400 and 13 of them are 16s. What is the probability of obtaining 413 or more 16s in 15,000 spins of a European wheel if it is not biased? Is there evidence that this particular wheel is biased?

13. In the same setting as the previous question, what is the probability of obtaining 602 or more 16s in 15,000 spins of a European wheel if it is not biased? Is there evidence that this particular wheel is biased?

14. If playing American roulette you bet $1 on red each time, what is the probability that you are ahead by at least $10 after 100 rounds?

15. In the same setting as the previous question, what is the probability that you will be ahead by at least $2 after 500 rounds?

16. [R] Corroborate the value of $P(Z \geq 160)$ when Z is binomial with $n = 300$ and $p = 18/38$ provided by the function `pbinom(159, 300, 18/38, lower.tail=FALSE)` in two ways:
    - By adding up the 141 terms involved in the sum.
    - Using a simulation.

17. [R] Modify the simulation of the Labouchère system to estimate the probability of going bankrupt if your bank is $200 and your list consists of 20 elements, each corresponding to $10 and you are making even bets.

18. [R] Modify the simulation of an even bet in roulette to corroborate the calculation of the expected number of spins before going bankrupt if you *do not* reinvest your winnings.

8

Blackjack

Blackjack (BJ) is a popular card game that has been depicted in movies such as the 2008 film *21*. Blackjack has become popular in good measure because it is one of the few casino games that can potentially be *broken* (i.e., a strategy can be devised to minimize or even eliminate the house advantage).

8.1 Rules and Bets

Blackjack (also called *21*) is played using a standard (French-style) 52-card deck (see Figure 8.1). The objective of the game is very simple: players try to get a combination of cards that adds up to a number that is larger than the dealer's number but does not exceed 21. The value of the cards is as follows: numbered cards are worth their value in points, Jacks, Queens, and Kings are worth 10 points each, and Aces are worth either 1 or 11 points (whichever is more advantageous to the player). The suits of the cards play absolutely no role in the outcome of the game.

At the start of the round each player is dealt two cards face up, and the dealer (let's call her Alice) also gets two cards. Cards are usually dealt from a stack which is called the *shoe.* The shoe can contain between one and eight decks.

Unlike the players, the dealer receives one card face up for everyone to see, and one card face down (which is said to be *in the hole*). After receiving the hidden card, Alice, the dealer, checks to see if she has a *blackjack* (an Ace plus a 10-point card such as a Jack, Queen, King, or 10). If Alice has a blackjack, she reveals the second card and the game ends, with all players who do not have a blackjack losing and any player with a blackjack tying with the house.

If Alice does not have a blackjack, then each player takes a turn playing. If the current player (let's call her Julissa) gets a *natural* 21, then she has a blackjack and automatically wins (unless, as we discussed before, the dealer also had a blackjack, in which case they draw). While the regular payoff odds in blackjack are 1 to 1, the payoff odds for a natural are 3 to 2 (i.e., you get a profit of $3 for each $2 you bet). If Julissa does not get a blackjack, she has the opportunity

Probability, Decisions and Games: A Gentle Introduction using R, First Edition. Abel Rodríguez and Bruno Mendes.
© 2018 John Wiley & Sons, Inc. Published 2018 by John Wiley & Sons, Inc.
Companion website: www.wiley.com/go/Rodriguez/Probability_Decisions_and_Games

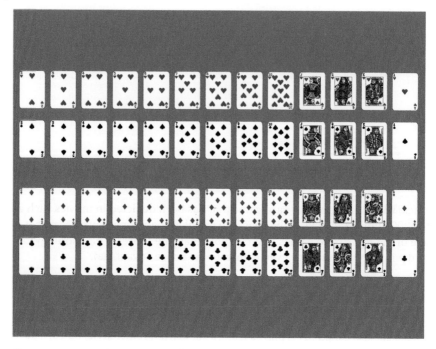

Figure 8.1 A 52-card French-style deck.

to draw as many additional cards (one by one) as she wants. More specifically, Julissa has the following options available to them:

- *Hit*: Draw an additional card. If the player's total, including the new card, goes above 21, the player *goes bust* and loses immediately, no matter whether the dealer later also goes bust or not.
- *Stay*: Stop drawing cards and wait for the dealer to play her hand.
- *Double-down* (only available as the first action of the hand): The player can double her bet if she agrees to stay after just one card.
- *Split* (only available as the first action of the hand): If the player gets a pair of cards with identical numbers, they can split the hand into two, placing an additional bet equal to the first one and drawing a card on each of the two hands. From then on, the player plays two independent hands simultaneously; the only restrictions are that, after a split, natural blackjacks are treated as regular 21s, and that further splits or double-downs are usually not allowed.
- *Surrender* (only available as the first action of the hand): Right after the dealer checks for blackjack, the player can surrender half of her bet and get back the other half. Surrendering is typically a bad option.
- *Insurance* (only available as the first action of the hand): If the dealer upside card is an Ace, the dealer might offer players the option to take insurance

against a blackjack before the dealer checks the hole card. The insurance bet becomes a side bet that the dealer has blackjack and is treated independently of the main wager. The payoff odds for this bet are 2 to 1.

Once all players have resolved their hands (either by going bust or staying), it is the turn of the dealer, who plays a fixed strategy. If Alice the dealer has not done so already (because of a blackjack), she shows the hole card. If the total is less than 17, she will hit until the number is higher than 17 or goes bust. If the dealer goes bust all players who stayed their game win. If the dealer did not go bust, then each player compares her number against the dealer's. If the player has a higher number, she wins; if the player has a smaller number, she loses. Finally, if the numbers are identical, the game is a draw and the player gets her bet back. In all these cases, the payoff odds are 1 to 1.

A popular variation of blackjack has the dealer hitting on a *soft* 17. A *soft number* is one made up of a combination of cards that includes an Ace that is counted for 11 points. For example, a combination of an Ace and a 6 is a soft 17, while a combination of a King, a 6, and an Ace counts as a *hard 17*. Other variations of the rules include *early surrender*, *re-splitting*, and *no doubling after splitting*. These variations are typically casino-specific and will not be discussed in this book.

8.2   Basic Strategy in Blackjack

Blackjack is popular in good part because it is possible for players to adopt a strategy that will minimize, or even eliminate, the house advantage. This is because (1) since the dealer plays a fixed strategy and she shows a face-up card, the player can adapt her strategy accordingly and (2) cards used in different rounds are typically dealt without replacement from a common (and finite!) deck, so the outcomes of different rounds are dependent. This is in clear contrast to the other games we have discussed so far (roulette, lotteries, craps) where outcomes from different rounds are independent from each other.

In order to devise a strategy for playing blackjack, let's consider first the probability associated with all possible dealer's hands. Since Alice must hit when her hand is under 17, there are 7 possible outcomes: 17, 18, 19, 20, 21, BJ, and bust (note that 21 means any combination of cards that adds up to 21 but are not blackjacks).

The probability of a blackjack is easy to compute,

$$P(\text{Blackjack})$$
$$= P\left(\begin{matrix}\text{First card}\\\text{is an A}\end{matrix}\right) \times P\left(\begin{matrix}\text{Second card is a}\\\text{10, J, Q or K}\end{matrix}\ \middle|\ \begin{matrix}\text{First card}\\\text{is an A}\end{matrix}\right)$$
$$+ P\left(\begin{matrix}\text{First card is a}\\\text{10, J, Q or K}\end{matrix}\right) \times P\left(\begin{matrix}\text{Second card}\\\text{is an A}\end{matrix}\ \middle|\ \begin{matrix}\text{First card is a}\\\text{10, J, Q or K}\end{matrix}\right).$$

In the case of a single-deck game this is

$$P(\text{Blackjack}) = \frac{4}{52} \times \frac{16}{51} + \frac{16}{52} \times \frac{4}{51} = \frac{2 \times 4 \times 16}{52 \times 51} \approx 0.04826546.$$

Single deck games, however, are relatively rare nowadays. In multiple deck games, the probability of different cards does not change much after a single card is removed. Hence, in that case, we can approximate

$$P(\text{Blackjack}) \approx \frac{4}{52} \times \frac{16}{52} + \frac{16}{52} \times \frac{4}{52} = \frac{2 \times 4 \times 16}{52 \times 52} \approx 0.04733728.$$

Note that the only difference between the two calculations is that the denominator for the probability that the second card is 52 when multiple decks are used (i.e., we assume sampling with replacement), instead of 51 when a single deck is used (in which case we are assuming sampling without replacement).

The probabilities for other outcomes can be quite complicated to obtain. Consider, for example, the probability of a 21 that is not a blackjack. This can only happen if three or more cards are drawn; there are many possible combinations that would lead to that outcome. For example,

- A 10-valued card, followed by a 6 and a 5.
- A 10-valued card, followed by a 5 and a 6.
- A 10-valued card, followed by a 4 and a 7.
- A 10-valued card, followed by a 4, a 2, and a 5.
- 9 followed by a 7 and then a 5.
- And so on…

A complete enumeration needs to be done carefully; for example, a 10-valued card followed by a 7 and then a 4 is not a combination that we should consider because it would never happen (once the 7 is drawn, the value of the hand is 17 and the dealer would stay). Since the number of combinations that we need to consider is rather large, we just present the results (for the multiple deck case) in Table 8.1. The following code can be used to corroborate the results using simulations:

Table 8.1 Probability of different hands assuming that the house stays on all 17s and that the game is being played with a large number of decks.

Result	17	18	19	20	21	BJ	Bust
Probability	0.145	0.140	0.134	0.180	0.073	0.047	0.282

```
> n = 100000
> cardvalues = rep(c(seq(1,10), rep(10,3)), each=4)
> outcome = rep(0,n)
> for(i in 1:n){
+    hand = sample(cardvalues, 2, replace=TRUE)
+    sw = TRUE
+    while(sw){
+      isace = (hand==1)
+      if(sum(isace)>0){
+        if(sum(hand[!isace]) + sum(isace) + 10 > 21){
+          handvalue = sum(hand[!isace]) + sum(isace)
+        }else{
+          handvalue = sum(hand[!isace]) + sum(isace) + 10
+        }
+      }else{
+        handvalue = sum(hand[!isace])
+      }
+      if(handvalue>=17){
+        sw = FALSE
+      }else{
+        hand = c(hand, sample(cardvalues, 1))
+      }
+    }
+    if(handvalue>21){
+      outcome[i]="Bust"
+    }else{
+      if(handvalue==21 & length(hand)==2){
+        outcome[i] = "BJ"
+      }else{
+        outcome[i] = handvalue
+      }
+    }
+ }
> round(table(outcome)/n, 3)

outcome
   17    18    19    20    21    BJ  Bust
0.147 0.140 0.132 0.180 0.073 0.047 0.281
```

Table 8.1 provides some interesting insights into blackjack strategy. For example, it shows that the dealer goes bust about once every four rounds. Since the player wins when the house goes bust as long they have not gone bust themselves, this suggests that it might be a good idea for the player to play defensively. However, the table does not make use of the knowledge provided by the face-up card. Indeed, note that about 30% of the cards in the deck are

10-valued cards. Therefore, the probability that the house goes bust is larger if the face-up card is a 6 than if it is a 10:

- If the face-up card is a 6, since the most likely scenario is that the hole card is a 10, the most likely total for the dealer's hand is 16. If that is the case, the dealer will have to draw a third card. If this happens, the most likely scenario is that the third card is a 6 or larger (probability 32/52), which leads to the dealer going bust. In this case, the player should play more defensively.
- If the face-up card is a 10, the probability that the dealer will be drawing a third card is small because they will only do it if the hole card is a 2, 3, 4, 5, or 6, which has a probability of 24/52 in a multideck game; so the probability that she will go bust is also relatively small. In this second case, the player should play more aggressively.

The previous discussion is formalized and generalized in Table 8.2, which shows the probability of different possible dealer hands conditional on the face-up card. As suggested earlier, the probability that the player goes bust is quite high if the face-up card is either 2, 3, 4, 5, or 6, but falls dramatically once the face-up card becomes a 7 or higher. Indeed, when the face-up card is a 7, the highest probability outcome is a 17, and for a 8 face-up card, the highest probability outcome is an 18, and so on. And, if the face-up card is an A, the probability that the dealer will go bust is very small. These probabilities suggest that it is a bad idea for the player to copy the house strategy. Instead, the following adaptive strategy is optimal:

Table 8.2 Probability of different hands assuming that the house stays on all 17s, conditional on the face-up card.

Face-up card	End hand						
	17	18	19	20	21	BJ	Bust
A	0.131	0.131	0.131	0.131	0.054	0.308	0.115
2	0.140	0.135	0.130	0.124	0.118	0.000	0.354
3	0.135	0.131	0.126	0.120	0.115	0.000	0.374
4	0.131	0.126	0.121	0.117	0.111	0.000	0.395
5	0.122	0.122	0.118	0.113	0.108	0.000	0.416
6	0.165	0.106	0.106	0.102	0.097	0.000	0.423
7	0.369	0.138	0.079	0.079	0.074	0.000	0.262
8	0.129	0.360	0.129	0.070	0.069	0.000	0.245
9	0.120	0.120	0.351	0.120	0.061	0.000	0228
10/J/Q/K	0.111	0.111	0.111	0.342	0.035	0.077	0.212

This table assumes that the game is being played with a large number of decks.

- If the bank's face-up card is either a 4, 5, or 6, the player should stay with any hand that is 12 or more, and hit otherwise.
- If the bank's face-up card is either a 2 or a 3, the players should stay with any number 13 or above.
- Against any other card, the player should stay with 17 or more and should hit otherwise.

The rationale for this strategy is directly linked to our earlier discussion. Since the bank has a very good chance of going bust if the card being shown is a 4, 5, or 6, the player should play very defensively, to the point of avoiding going bust at all cost (hence the strategy of staying with 12 or more). On the other hand, if the house shows a 7 or higher in the face-up, the probability of a good hand for the house is very high and the player should play more aggressively to try to get at least a 17 before staying.

A similar intuition works for the optimal splitting strategy (see Table 8.3). Note that splitting on a double 10 is never advantageous; this is so because the probability that the dealer will beat a 20 is very small no matter what the face-up card is. Similarly, splitting with a double 7 is advantageous only if the dealer shows a 7 or smaller number (for larger face-up cards, the probability that the player goes bust or gets a number that is 17 or less is very high, while the probability that the dealer will get a number that is 18 or more is also relatively high).

Table 8.3 Optimal splitting strategy.

	Dealer's face-up card									
Player's card	2	3	4	5	6	7	8	9	10	A
A–A	S	S	S	S	S	S	S	S	S	
10–10										
9–9	S	S	S	S	S		S	S		
8–8	S	S	S	S	S	S	S	S		
7–7	S	S	S	S	S	S				
6–6	(S)	S	S	S	S					
5–5										
4–4				(S)	(S)					
3–3	(S)	(S)	S	S	S	S				
2–2	(S)	(S)	S	S	S	S				

S indicates that splitting is advantageous, while (S) indicates situations in which splitting is advantageous only if doubling down is allowed. This strategy assumes that the game is being played with a large number of decks and that the dealer stays at all 17s.

8.3 A Gambling System that Works: Card Counting

A key feature of blackjack is that the deck (or decks) of cards is not reshuffled after each hand. Instead, multiple rounds are dealt continuously from the same deck. This can induce some wild variations in the probabilities of different hands. This phenomenon is more pronounced when playing with a single deck, but it can still be exploited in multi-deck games to create a gambling system.

To see how much the probabilities of different outcomes can change in a single-deck game, consider a situation where all Aces, 2s, 3s, 4s, 5s, and 6s have been removed from a deck (so, there are 28 cards left in the deck; 16 ten-valued cards, four 7s, four 8s, and four 9s). In this case, the probability of different hands is relatively easy to compute. For example, the probability that the dealer gets a 21 is simply the probability of getting three 7s in a row (there are no other combinations available with the cards left in the deck), that is,

$$P(21) = \underbrace{\frac{4}{28}}_{\substack{\text{Probability that the} \\ \text{first card is a 7}}} \times \underbrace{\frac{3}{27}}_{\substack{\text{Probability that the} \\ \text{second card is a 7} \\ \text{given first is a 7}}} \times \underbrace{\frac{2}{26}}_{\substack{\text{Probability that the} \\ \text{third card is a 7} \\ \text{given the first two are 7s}}} \approx 0.00122.$$

Similarly, the probability that the dealer gets a 17 corresponds to the probability of 4 different sequences: first a 7, then a 10, or first a 10, then a 7, or first an 8, then a 9, or first a 9, then an 8. Therefore,

$$P(17) = \frac{4}{28} \times \frac{16}{27} + \frac{16}{28} \times \frac{4}{27} + \frac{4}{28} \times \frac{4}{27} + \frac{4}{28} \times \frac{4}{27} \approx 0.21164.$$

Table 8.4 summarizes the probabilities for all cases, which are very different from those in Table 8.1 because a number of cards are not in play anymore. The values in Table 8.4 can be corroborated using the following R code (note that this version of the code, unlike the one presented in Section 8.2, uses sampling without replacement):

```
> n = 100000
> cardvalues = rep(c(seq(7,10), rep(10,3)), each=4)
> outcome = rep(0,n)
> for(i in 1:n){
+    shuffleddeck = sample(cardvalues,
+                          length(cardvalues), replace=FALSE)
+    hand       = shuffleddeck[1:2]
+    currentcard = 3
+    sw = TRUE
+    while(sw){
+      isace = (hand==1)
+      if(sum(isace)>0){
+        if(sum(hand[!isace]) + sum(isace) + 10 > 21){
+          handvalue = sum(hand[!isace]) + sum(isace)
```

```
+        }else{
+            handvalue = sum(hand[!isace]) + sum(isace) + 10
+        }
+    }else{
+        handvalue = sum(hand[!isace])
+    }
+    if(handvalue>=17){
+        sw = FALSE
+    }else{
+        hand = c(hand, shuffleddeck[currentcard])
+        currentcard = currentcard + 1
+    }
+ }
+ if(handvalue>21){
+     outcome[i]="Bust"
+ }else{
+     if(handvalue==21 & length(hand)==2){
+         outcome[i] = "BJ"
+     }else{
+         outcome[i] = handvalue
+     }
+ }
+ }
> round(table(outcome)/n, 3)

outcome
  17     18     19     20     21   Bust
0.211  0.184  0.171  0.316  0.001  0.116
```

Table 8.4 Probability of different hands assuming that the house stays on all 17s and that the game is being played with a single deck where all Aces, 2s, 3s, 4s, 5s, and 6s have been removed.

Result	17	18	19	20	21	BJ	Bust
Probability	0.212	0.186	0.170	0.317	0.001	0.000	0.115

A similar approach can be used to compute the probability of each outcome conditional on the face-up card (this is analogous to the calculation behind Table 8.2). For example, if the face-up card is a 7, the only second cards not leading to an automatic bust are a 10 (which leads to 17) and a 7 (which leads to drawing a third card since the running total is 14). If the second card is a 7, then the only way the dealer will not go bust is with a third 7 (which leads to a total of 21). Thus,

$$P\left(17 \,\middle|\, \begin{array}{l} \text{First card is a 7 } \underline{and} \text{ A,} \\ 2, 3, 4, 5, 6 \text{ removed} \end{array}\right)$$

$$= P\left(\begin{array}{c|c} \text{Second card} & \text{First card is a 7 \underline{and} A,} \\ \text{is 10-valued} & \text{2, 3, 4, 5, 6 removed} \end{array}\right) = \frac{16}{27} \approx 0.59259,$$

$$P\left(21 \; \middle| \; \begin{array}{c} \text{First card is a 7 \underline{and} A,} \\ \text{2, 3, 4, 5, 6 removed} \end{array}\right) = \frac{3}{27} \times \frac{2}{26} \approx 0.008547,$$

while

$$P(18 \mid \text{First card is a 7 \underline{and} A, 2, 3, 4, 5, 6 removed}) = 0,$$

$$P(19 \mid \text{First card is a 7 \underline{and} A, 2, 3, 4, 5, 6 removed}) = 0,$$

$$P(20 \mid \text{First card is a 7 \underline{and} A, 2, 3, 4, 5, 6 removed}) = 0,$$

and

$$P\left(\text{Bust} \; \middle| \; \begin{array}{c} \text{First card is a 7 \underline{and} A,} \\ \text{2, 3, 4, 5, 6 removed} \end{array}\right) = 1 - \overbrace{\left(\frac{16}{27} + \frac{3}{27} \times \frac{2}{26}\right)}^{\substack{\text{Probability dealer} \\ \text{does } not \text{ go bust}}} \approx 0.39886.$$

After rounding to four decimal places, these results correspond to the values in Table 8.5. The rest of the table can be obtained in a similar way.

Because the probability of winning a hand depends greatly on which cards are left in the deck, we can reduce (and even eliminate) the house advantage by increasing the amount of our bet when the content of the decks drives the odds of winning in our favor and decreasing it when the odds are against us.

Card counting is a simple mechanism that can be used to keep track of the cards that have appeared in previous rounds dealt since reshuffling, and then adapting the size of the bets in order to exploit situations that are favorable

Table 8.5 Probability of different hands assuming that the house stays on all 17s, conditional on the face-up card.

Face-up card	Dealer's final hand						
	17	18	19	20	21	BJ	Bust
7	0.593	0.000	0.000	0.000	0.009	0.000	0.399
8	0.148	0.593	0.000	0.000	0.000	0.000	0.259
9	0.148	0.111	0.593	0.000	0.000	0.000	0.148
10/J/Q/K	0.148	0.148	0.148	0.556	0.000	0.000	0.000

This table assumes that the game is being played with a single deck and all the Aces, 2s, 3s, 4s, 5s, and 6s have been removed from a deck.

to the player. *Counting systems* are based on the fact that high cards (especially Aces and 10s) benefit the player more than the dealer, while the low cards (especially 4s, 5s, and 6s) help the dealer while hurting the player. Indeed, a high concentration of Aces and 10s in the deck increase the player's chances of hitting a natural blackjack, which pays out 3:2 (unless the dealer also has blackjack). Also, when the deck has a high concentration of 10s, players have a better chance of winning when doubling. On the other hand, low cards benefit the dealer, since according to blackjack rules the dealer must hit *stiff* hands (i.e., hands totaling 12–16) while the player has the option to hit or stand. Consequently, a dealer holding a stiff hand will bust every time if the next card drawn is a 10.

A number of counting systems have been devised. All of them start the count at zero when a deck is freshly shuffled and increase/decrease the count according to the cards that are played. The simplest point assignment for the cards that is used in practice is the low-high count system:

- 2s, 3s, 4s, 5s, 6s are assigned a value of +1.
- 10-Valued cards and A are assigned a value of −1.
- All other numbers (7, 8, and 9) are assigned a value of 0.

When the count is high, it signals a lot of high cards left in the deck which means that it is easier for the dealer to go bust; in this situation, you should increase your bets because it is more likely for you to win. Different experts suggest different thresholds for increasing the bet, one possible option (for single-deck games) is the following:

- If your count is less than or equal to +1, bet the table minimum.
- If your count is between +2 and +3, double your minimum bet.
- If your count is between +4 and +5, triple your minimum bet.
- If your count is between +6 and +7, quadruple your minimum bet.
- If your count is +8 or more, quintuple your minimum bet.

8.4 Exercises

1. Explain why, in single-deck blackjack, there is an advantage to the player when a large number of high cards are in the deck and the mid-valued cards are out.

2. Why is it usually a bad idea to surrender in blackjack?

3. What actions can casinos take to reduce the advantage of card counters in blackjack? Explain the logic behind these actions.

4. You are playing blackjack from a single deck, and you are the only player on the table. Your hand is K–8 and the dealer shows a 9. If you know that all Aces, 2s, 3s, 4s, 5s, and 6s are out of the deck (but all other cards are still in), what is the probability that you will win the hand if you stay?

5. What's the probability of winning an insurance bet? What is your expected profit in a game where you take an insurance bet?

6. Consider the same situation as the previous question; what is the probability that you will win when the dealer's hand is showing a 7.

7. Consider the same situation as the previous question; what is the probability that you will win when the dealer's hand is showing a 8.

8. When playing with a deck where all Aces, 2s, 3s, 4s, 5s, and 6s are out of the deck (but all other cards are still in) what is the probability of the dealer getting a 17 when her hand is showing a 10. Explain your reasoning carefully.

9. In the same situation as the previous question, what is the probability of the dealer getting an 18 when her hand is showing a 9. Justify your answer.

10. Still in the same situation as two previous questions, what is the probability of the dealer going bust when her hand is showing a 9. Justify your answer.

11. When using a continuous shuffling machine (CSM), cards just used are placed back into the deck of cards at the end of each hand. The machine works in such a way that any of the cards just played has a chance of coming up in the next hand (unlike in regular games, where cards that are discarded are not reintroduced for a while). How does a CSM affect the effectiveness of the basic strategy in blackjack? How does it affect the effectiveness of card counting?

12. Carry out the computations required to complete Table 8.5.

13. [R] Modify the simulation presented in this chapter to re-compute the probabilities in Table 8.1 for a single-deck game in which cards are sampled without replacement.

14. **[R]** Write code to evaluate the house advantage under the basic black-jack strategy discussed in Section 8.2 assuming no insurance, surrender, splitting, or doubling down is allowed.

15. **[R]** Modify the code from the previous exercise to compute the house advantage if the player copies the house strategy. Compare your results against those from the previous exercise.

9

Poker

Poker is a very popular type of card game played not only in casinos but also among friends. One of its variants, called *Texas Hold'em*, has become particularly popular since sports channels such as ESPN started showcasing championships.

Poker is different from the other games we have discussed so far in that the players compete against each other rather than against the casino. Therefore, even though poker has features that are akin to other random games, it is also a game of strategy. In this section, we will discuss the random aspects of the game and will delay the discussion of its strategic aspects until Chapter 11.

9.1   Basic Rules

Just like blackjack, poker is played using a French-style 52-card deck (recall Figure 8.1). Each player is dealt a certain number of cards either face down or face up. In addition, some community cards (which are shared by all players) might be dealt. The winner of each game is the person with the highest five-card hand; the way in which the hand is constructed using the player's and community cards depends on the variant of the game being played (more on this later).

Hands are ranked primarily by their type (see Table 9.1 and Figure 9.1). Card numbers are used only to differentiate among hands of the same type. For this purpose, the cards are ordered 2, 3, 4, 5, 6, 7, 8, 9, 10, J, Q, K, A. Suits by themselves do not typically play a role in defining the value of a hand, except for helping determine if you have a flush. To understand how hands are compared, let's consider a few examples.

1. Assume your hand is 2♢ 2♣ 3♠ 7♡ 10♡ and your opponent's hand is A♢ Q♣ 4♣ 8♣ 10♢. You have one pair of 2s, while your opponent has an Ace high card. You win the game because a pair beats a single higher card.
2. Assume now that your hand is 2♢ 2♣ Q♠ Q♡ Q♣, while your opponent's hand is A♠ A♣ 3♠ 3♣ 3♡. Both of you have full houses, so we need to look

Probability, Decisions and Games: A Gentle Introduction using R, First Edition. Abel Rodríguez and Bruno Mendes.
© 2018 John Wiley & Sons, Inc. Published 2018 by John Wiley & Sons, Inc.
Companion website: www.wiley.com/go/Rodriguez/Probability_Decisions_and_Games

Table 9.1 List of poker hands.

Rank	Name	Description
1	Royal flush	The hand contains the A, K, Q, J and 10 of the same suit.
2	Straight flush	The hand contains five cards of the same suit with consecutive values. A can come before a 2, but not after K (as the hand would be a Royal Flush).
3	Four of a kind (poker)	The hand contains four cards of the same number (one for each suit).
4	Full house	The hand contains three cards of one number and two cards of a different number.
5	Flush	The hand contains five cards of the same suit, but not a Straight flush.
6	Straight	The hand contains five cards with consecutive number values that are not a Straight flush.
7	Three of a kind	The hand contains three cards of the same number and is not a full house or a poker.
8	Two pairs	The hand contains two pairs, each of a different number.
9	One pair	The hand contains two cards of the same number and is not a full house or a poker.
10	Highest card	The hand is not any of the above.

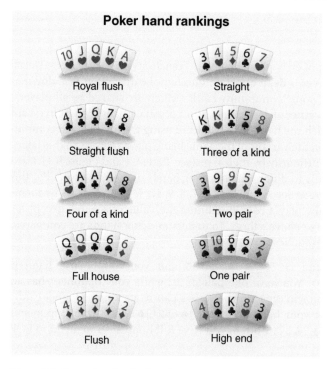

Figure 9.1 Examples of poker hands.

at the number associated with the cards in order to compare hands. We first compare the numbers associated with the three of a kind. Since you have Qs and your opponent has 3s, you win the hand.

3. Finally, assume that your hand is A♣ K♣ Q♣ J♣ 10♣ and your opponent's hand is A♡ K♡ Q♡ J♡ 10♡. Both of you have (Royal) Straight flushes, so the hand is a tie and the pot is split among the players.

In poker, betting rounds are usually interspaced between rounds of dealing cards (the specific details depend, again, on the variant you are playing). During these betting rounds, players take turns deciding whether to withdraw from the game (*fold*), increase their bets (*raise*), or match a raise by another player (*call*).

9.2   Variants of Poker

There is a large number of variants of poker, which differ mainly on how the player's hands are formed. Three of them are particularly popular nowadays.

In *draw poker*, a complete hand is dealt to each player, face down and, after betting, players are allowed to improve their hand by discarding unwanted cards and being dealt new ones. After the card exchange, a second round of betting ensues. *Five-card draw* is the most popular version of draw poker.

In *stud poker*, cards are dealt in a prearranged combination of face-down and face-up rounds, or *streets*, with a round of betting following each dealing. The most popular stud variant today, seven-card stud, deals seven cards to each player (three face down, four face up) from which they must make the best possible five-card hand.

Community card poker is similar to stud poker in that cards are dealt to the players in a combination of face-down and face-up cards. However, in community card poker, the face-up cards are shared by all the players. Players are dealt an incomplete hand of face-down cards, and then a number of face-up community cards are dealt to the center of the table, each of which can be used by one or more of the players to make a five-card hand. *Texas Hold'em* is the best-known community card poker. In Texas Hold'em, each player receives two face-down cards. In addition, five face-up community cards are shared by all players. These cards are dealt in the following order:

1. First, the two face-down cards are dealt to each player, followed by a round of betting.
2. The first three community cards are revealed (*the flop*), followed by a second round of betting.
3. The fourth community card is revealed (*the turn*), followed by a third round of betting.
4. The final community card is revealed (*the river*), followed by a fourth and last round of betting.

9.3 Additional Rules

In addition to the voluntary bets made by each player during the game, *forced bets* are often employed to create incentives for the players to wager even when hands are bad. *Blinds* are a forced bet placed by one or more players that is made before the cards are dealt; these are used very often in draw and community poker. The *bring-in* is another forced bet that occurs after the cards are initially dealt, but before any other action is taken. Bring-ins are common in stud poker, and it is required from the player with the worst set of open cards.

All in bets are another special type of wager. If you are faced with a bet you cannot match for the lack of sufficient funds, you may bet the remainder of your stack and declare yourself all in. You may now hold onto your cards for the remainder of the round as if you had called every bet, but you may not win any more money from any player above the amount of your stack.

9.4 Probabilities of Hands in Draw Poker

It is particularly easy to compute the probabilities of different hands in draw poker because players have no information about other players' cards. We start by computing the probabilities of getting the different types of hands with the first five cards. To compute these probabilities, recall that a hand of poker consists of five cards drawn randomly without replacement from a single, well-shuffled deck of 52 cards. Therefore, all cards have the same probability of appearing in your hand, cards cannot repeat themselves (you can get two Qs, like Q$\diamond$ and Q$\spadesuit$, but your hand cannot have two Q$\spadesuit$), and the order in which the cards are arranged is irrelevant. This implies that we are dealing with an equiprobable space where

$$P(\text{Type of hand}) = \frac{\text{Number of hands consistent with the goal}}{\text{Total number of possible poker hands}}.$$

Because we are drawing cards without replacement and order does not matter, the total number of possible hands of poker is

$$\text{Total number of possible poker hands} = \binom{52}{5} = \frac{52!}{5! \times 47!}$$

$$= \frac{52 \times 51 \times 50 \times 49 \times 48}{5 \times 4 \times 3 \times 2 \times 1} = 2,598,960.$$

(Recall Chapter 4.)

Now we only need to compute the number of hands that correspond to each one of the named hands mentioned in Table 9.1. For the *Royal flush* and *Straight flush*, note that for each of the four suits there are 10 different possible five-card sequences:

| A,2,3,4,5 | 2,3,4,5,6 | 3,4,5,6,7 | 4,5,6,7,8 | 5,6,7,8,9 |
| 6,7,8,9,10 | 7,8,9,10,J | 8,9,10,J,Q | 9,10,J,Q,K | 10,J,Q,K,A |

The very last one corresponds to a Royal flush, while the other nine are regular Straight flushes. Since there are four suits in the deck, which means that there are four combinations of cards that yield a Royal flush and 36 that yield a Straight flush. Therefore,

$$P(\text{Royal flush}) = \frac{4}{2,598,960} \approx 0.00000154,$$

$$P(\text{Straight flush}) = \frac{36}{2,598,960} \approx 0.00001385.$$

For *four of a kind*, we can split the problem into two parts: first, we figure out how many sets of four cards with the same number can come up, and then we figure out how many options are available for the fifth card in the hand. Since there are 13 possible numbers, the number of possible sets of four cards is very easy: there are 13 of them. On the other hand, the fifth card could be any of the 48 cards left in the deck. Therefore,

$$P(\text{Four of a kind}) = \frac{13 \times 48}{2,598,960} \approx 0.0002401.$$

Let's look now at the probability of a *full house*. As before, we break the problem into two parts; we first compute the number of possible trios that can come up, and then we compute the number of pairs. For the number of trios, we have 13 possible options for the number associated with the trio, and we have $\binom{4}{3} = \frac{4!}{3! \times 1!} = 4$ options for the combination of suits associated with these three cards (remember that you have four cards with each number, one for each suit, and we need to pick three of them). Therefore, the number of trios is $13 \times 4 = 52$. A similar reasoning can be used for the number of pairs; there are now 12 possible numbers that you could use for the pair (the pair has to have a number that is different from the trio, leaving you 12 rather than 13 options), and there are $\binom{4}{2} = \frac{4!}{2! \times 2!} = 6$ combinations of suits for that number, for a total of $12 \times 6 = 72$ distinct pairs. Therefore,

$$P(\text{Full house}) = \frac{13 \times 4 \times 12 \times 6}{2,598,960} \approx 0.001440576.$$

Next we consider the probability of a *flush*, which is composed of five cards of the same suit. Since there are four possible suits and, for a given suit, there are $\binom{13}{5} = \frac{13!}{5! \times 8!} = 1287$ sets of five cards (recall that each suit has 13 different cards), there are $4 \times 1287 = 5148$ such hands. However, this includes the Straight flushes, which should not be included in the count and need to be subtracted (recall that there are 10 Straight flushes for each suit, including Royal flushes). Therefore, the probability of a flush is

$$P(\text{Flush}) = \frac{4 \times (1287 - 10)}{2{,}598{,}960} \approx 0.0019654.$$

The reasoning for a *straight* is very similar to that for the flush. Recall that there are 10 different straight sequences:

A,2,3,4,5   2,3,4,5,6   3,4,5,6,7   4,5,6,7,8   5,6,7,8,9

6,7,8,9,10   7,8,9,10,J   8,9,10,J,Q   9,10,J,Q,K   10,J,Q,K,A

In principle, there are four options for the suit of the first card, four options for the suit of the second, and so on. Consequently, there are $4^5 = 1024$ combinations of suits for each of the 10 straight sequences. However, this number again includes the Straight flushes, so we need to subtract them. Therefore,

$$P(\text{Straight}) = \frac{10 \times (1024 - 4)}{2{,}598{,}960} \approx 0.003924647.$$

For the probability of *three of a kind*, note that, as with the full house, there are 13 options for the number and $\binom{4}{3} = \frac{4!}{3! \times 1!} = 4$ choices for the suits of these three cards. For the fourth and fifth cards of the hand, they might be of any suit, but their numbers need to be different from each other, and different from the number used for the trio (otherwise, you would have a full house or four of a kind). Accordingly, there are $4 \times 4 = 16$ options for the suits of the remaining two cards and $\binom{12}{2} = \frac{12!}{10! \times 2!} = 66$ choices for their numbers, leading to:

$$P(\text{Three of a kind}) = \frac{13 \times 4 \times 16 \times 66}{2{,}598{,}960} \approx 0.02112845.$$

For *two pairs* the calculation is very similar. First, we need to pick two numbers out of 13 possible options (remember that the numbers in both pairs need to be from different suits or you have four of a kind), which yields $\binom{13}{2} = \frac{13!}{11! \times 2!} = 78$ options. Then, we need to choose the suits for each of the pairs (there are $\binom{4}{2} = \frac{4!}{2! \times 2!} = 6$ options for the suits). Finally, we need to look at the fifth card, which can be any card out of the remaining 44 (you need to exclude the eight cards that correspond to any of the numbers in the pairs, or you would have a full house instead of two pairs). Therefore,

$$P(\text{Two pairs}) = \frac{78 \times 6 \times 6 \times 44}{2{,}598{,}960} \approx 0.04753902.$$

Finally, the probability of a *single pair* is

$$P(\text{Single pair}) = \frac{13 \times 6 \times 64 \times 220}{2{,}598{,}960} \approx 0.4225.$$

This results from realizing that there are 13 options for the number of the pair, $\binom{4}{2} = \frac{4!}{2! \times 2!} = 6$ options for the suits of the two cards in the pair, $4 \times 4 \times 4 = 64$

options for the suit of the remaining three remaining cards, and $\binom{12}{3} = \frac{12!}{9! \times 3!} =$ 220 options for the number of the three remaining cards.

The calculations we just discussed can be easily corroborated using simulations. For example, the probability of three of a kind can be approximated as

```
> n = 100000
> cardnumbers = rep(c(seq(2,10), "J", "Q", "K", "A"), 4)
> cardsuits   = rep(c("S", "C", "H", "D"), each=13)
> isthreeofakind = rep(FALSE,n)
> for(i in 1:n){
+     carddealt = sample(seq(1,52), 5, replace=FALSE)
+     yourcardnumbers = cardnumbers[carddealt]
+     yourcardsuits   = cardsuits[carddealt]
+     x = sort(table(yourcardnumbers))
+     if(length(x)==3 & x[1]==1 & x[2]==1){
+         isthreeofakind[i] = TRUE
+     }
+ }
> sum(isthreeofakind)/n

[1] 0.02221
```

Similarly, for the probability of two pairs,

```
> n = 100000
> cardnumbers = rep(c(seq(2,10), "J", "Q", "K", "A"), 4)
> cardsuits   = rep(c("S", "C", "H", "D"), each=13)
> istwopairs = rep(FALSE,n)
> for(i in 1:n){
+     carddealt = sample(seq(1,52), 5, replace=FALSE)
+     yourcardnumbers = cardnumbers[carddealt]
+     yourcardsuits   = cardsuits[carddealt]
+     x = sort(table(yourcardnumbers))
+     x
+     if(length(x)==3 & x[1]==1 & x[2]==2){
+         istwopairs[i] = TRUE
+     }
+ }
> sum(istwopairs)/n

[1] 0.04832
```

9.4.1 The Effect of Card Substitutions

We consider now how the probabilities of the different hands are affected when you are allowed to replace some cards in your hand. For example, consider the probability of getting a Straight flush if you are allowed to exchange up to one card. In this case, we can reformulate the problem as the probability of getting a Straight flush if you are dealt six rather than five cards from the deck. Indeed, the extra card might or might not be actually drawn depending on whether you get the Straight flush in the first hand, but the calculation is unaffected since you only exchange cards if you need to.

In this case, the total number of possible hands is $\binom{52}{6} = 20{,}358{,}520$. To compute the number of hands consistent with the desired outcome, note that five of the cards need to correspond to the desired outcome (a Straight flush) therefore, as before, there are 36 possible options for the first five cards. On the other hand, the sixth card could potentially be any other card in the deck (so there are 47 options left). Hence,

$$P\left(\begin{array}{c}\text{Straight flush if we are allowed}\\\text{to exchange up to one card}\end{array}\right) = \frac{36 \times 47}{20{,}358{,}520}$$

$$= \frac{6 \times 36}{2{,}598{,}960} \approx 0.0000831.$$

Hence, by allowing the player to exchange one card, the probability of a Straight flush, although still small, is six times higher than before!

9.5 Probabilities of Hands in Texas Hold'em

Texas Hold'em is nowadays the most widely played variant of poker. The use of multiple community cards offers more opportunities to bet than draw poker (allowing for more strategic play) and makes the game less predictable. Indeed, as the flop, turn, and river are revealed, the probabilities that each player wins can change dramatically. In televised games, this is exploited for dramatic effect by showing the cards held by the players along with the community cards and the changing probabilities that each of the player wins.

Recall that, in Texas Hold'em, the player is first dealt two face-down cards (sometimes called *hole* or *pocket* cards), followed by a first round of betting. The number of possible pocket hands is relatively small,

$$\text{Number of pocket hands} = \binom{52}{2} = 1326,$$

which is the number of ways in which a pair of cards can be drawn from a deck of 52 cards without replacement, and the order of the two cards is not important. Computing the probability of the different pocket hands is Straight forward. For example, we can compute the probability of getting a *pocket pair* (i.e., the probability that the two pocket cards form a pair) as

$$P(\text{Pocket pair}) = \frac{13 \times 6}{1326} \approx 0.0588,$$

Table 9.2 List of opponent's poker hands that can beat our two-pair.

	Opponent's winning hand	Opponent's hidden cards	Probability
Two pairs	Two Qs and two 8s	Q & any except 2s, 8s, Qs, or Ks	$\frac{204}{1980}$
	Two Ks and two 8s	K & any except 2s, 8s, Qs, or Ks	$\frac{204}{1980}$
	Two As and two 8s	A & A	$\frac{12}{1980}$
	Two Qs and two Ks	Q & K	$\frac{18}{1980}$
	Two Qs and two 2s	Q & 2	$\frac{18}{1980}$
	Two Ks and two 2s	K & 2	$\frac{18}{1980}$
Three of a kind	Three 8s	8 & any except 2s, 8s, Qs, or Ks	$\frac{136}{1980}$
Full house	Three 2s and two 8s	2 & 2	$\frac{6}{1980}$
	Three Ks and two 8s	K & K	$\frac{6}{1980}$
	Three Qs and two 8s	Q & Q	$\frac{6}{1980}$
	Three 8s and two 2s	8 & 2	$\frac{12}{1980}$
	Three 8s and two Ks	8 & K	$\frac{12}{1980}$
	Three 8s and two Qs	8 & Q	$\frac{12}{1980}$
Four of a kind	Four 8s	8 & 8	$\frac{2}{1980}$

where the numerator comes from the fact that we have 13 different options for the number of the pair, and $\binom{4}{2} = 6$ possible combinations of suits for the pair.

Computing the probability of winning a hand in Texas Hold'em requires that we condition on the community cards that have been revealed. For example, assume that your hand is J♣ J♠ and the face-up cards are 2◊ K♠ 8♡ Q♣ 8♣. In that case you have two pairs, one of which is shared by all players (the pair of 8s). The winning hands for an opponent (let's call him Malik) are shown in Table 9.2. We now proceed to compute the probabilities associated with each of these hands.

If Malik has two Qs and two 8s, he will beat your two Js and two 8s. Malik can have this hand if he holds a Q and any other card (excluding 2s, 8s, Qs, or Ks because these would form a three of kind or a full house, which will be considered later). Hence,

$$P\begin{pmatrix}\text{Opponent's winning} \\ \text{hand with two pairs} \\ \text{that include a Q}\end{pmatrix} = P\left(\begin{array}{c}\text{(Q \underline{and} any card but 2, 8, Q, or K) \underline{or}} \\ \text{(any card but 2, 8, Q, or K \underline{and} Q)}\end{array}\right).$$

Now the probability that Malik gets a Q and any other card in his hand is

$$P(\text{Q \underline{and} any card but 2, 8, Q, or K}) =$$

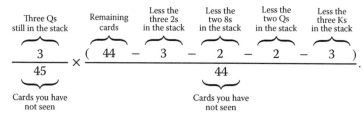

	Three Qs still in the stack	Remaining cards	Less the three 2s in the stack	Less the two 8s in the stack	Less the two Qs in the stack	Less the three Ks in the stack

$$\underbrace{\frac{3}{45}}_{\substack{\text{Cards you have}\\\text{not seen}}} \times \frac{\left(\overbrace{44} - \overbrace{3} - \overbrace{2} - \overbrace{2} - \overbrace{3}\right)}{\underbrace{44}_{\substack{\text{Cards you have}\\\text{not seen}}}}.$$

The calculation for $P(\text{any card but 2, 8, Q, or K \underline{and} Q})$ is identical. Therefore,

$$P\begin{pmatrix}\text{Opponent's winning} \\ \text{hand with two pairs} \\ \text{that include a Q}\end{pmatrix} = \frac{3 \times 34}{1980} + \frac{34 \times 3}{1980} = \frac{204}{1980}.$$

A second way in which you can lose is if Malik has two Ks and two 8s. This can happen if he holds a K and any other card (except 2s, 8s, Qs, or Ks, which would produce a stronger hand than two pair and will be considered below). This calculation goes exactly like the calculation for the previous situation. The probability is therefore $\frac{204}{1980}$.

Another possibility is for Malik to have two As and two 8s. This hand can arise if he holds two As in his hand. The probability of this occurring is

$$P(\text{A \underline{and} A}) = \frac{4}{45} \times \frac{3}{44} = \frac{12}{1980}.$$

Your opponent's two pairs that beat your two pairs are two Qs and two Ks, two Qs and two 2s, and finally two Ks and two 2s. Each of those situations can be realized if Malik holds a Q and a K, a Q and a 2, or a K and a 2, respectively. The probabilities for these three options are calculated the same way, so we focus on the probability of Malik holding a Q and a K:

$$P(\text{Q \underline{and} K}) = P(\text{Q \underline{and} K}) + P(\text{K \underline{and} Q})$$
$$= 2 \times \left(\frac{3}{45} \times \frac{3}{44}\right) = \frac{18}{1980}.$$

Malik can also beat you with three of a kind. This can only happen if Malik holds an 8 and any other card (except the usual 2s, 8s, Qs, or Ks) in his hand. Indeed, notice that if Malik holds two 2s, it will produce not a trio but a full house if you consider the two 8s in the community cards. For the same reason, if Malik holds two Qs or two Ks, they will not count as three of a kind, but as full houses. The probability of three of a kind is therefore,

$$P\left(\begin{array}{c}\text{8 \underline{and} any card}\\\text{but 2, 8, Q, or K}\end{array}\right) = P\left(\begin{array}{c}\text{8 \underline{and} any card}\\\text{but 2, 8, Q, or K}\end{array}\right) + P\left(\begin{array}{c}\text{Any card but 2,}\\\text{8, Q or K \underline{and} 8}\end{array}\right)$$

$$= 2 \times \left(\frac{2}{45} \times \frac{44 - 3 - 1 - 3 - 3}{44}\right)$$

$$= 2 \times \frac{68}{1980} = \frac{136}{1980}.$$

We now consider the next stronger hand, a full house. This can be realized in six different ways:

1. Malik holds two 2s, forming a set of three 2s and two 8s.
2. Malik holds two Ks, resulting in a set of three Ks and two 8s.
3. He can hold two Qs, forming a set of three Qs and two 8s.
4. If he holds an 8 and a 2, his hand would be three 8s and two 2s.
5. He can also hold an 8 and a K, resulting in three 8s and two Ks.
6. Finally, if Malik holds an 8 and a Q, he will have three 8s and two Qs.

The probability for the first, second, and third possibilities are calculated in the same way. For example,

$$P(2 \text{ \underline{and} } 2) = \frac{3}{45} \times \frac{2}{44} = \frac{6}{1980}.$$

On the other hand, the probability of holding an 8 and a 2 is

$$P(\text{Opponent holds an 8 and a 2}) = P(8 \text{ \underline{and} } 2) + P(2 \text{ \underline{and} } 8)$$

$$= \frac{2}{45} \times \frac{3}{44} + \frac{3}{45} \times \frac{2}{44} = \frac{12}{1980},$$

which goes exactly the same way for the probability of obtaining a Q and a 8.

There is just one more way Malik can beat your two pairs; if he has two 8s in his hands, he will be able to form a poker. The probability of this happening is

$$P(8 \text{ \underline{and} } 8) = \frac{2}{45} \times \frac{1}{44} = \frac{2}{1980}.$$

Once all the cases have been considered, the probability that you lose can be calculated as the sum of all the probabilities in the last column of Table 9.2:

$$P(\text{Opponent wins}) = \frac{666}{1980} \approx 0.3515152,$$

and the probability of a tie is the probability that Malik has the two Js left in the deck, that is,

$$P(\text{Tie}) = \frac{1}{\binom{45}{2}} = \frac{2}{1980} \approx 0.001.$$

Given that your probability of winning is relatively large (around 66%), you would probably do well to raise in the last round of betting!

9.6 Exercises

1. In traditional *draw poker* (where you are given five cards that are unknown to your opponents), what is the probability that you will be dealt four of a kind in the first hand? If you were allowed to change one card, what would be the probability?

2. What is the probability that you will be dealt a flush in the first hand in traditional draw poker? If you were allowed to change one card, what would be the probability?

3. Considering your calculations in the previous two questions, is it beneficial to the player to be allowed to swap cards?

4. In traditional draw poker, if we are allowed to exchange one card and the current hand is 7◇10♣8♡5◇K◇, what is the probability that get a pair if you switch only one card? What is the probability if you decide to exchange four cards and only keep the highest card in your hand (the K◇)?

5. You are playing five-card stud poker without bring-in and only one opponent is left. You show 2◇3♠Q♠Q♡ and your hidden card is A♡. Your opponent's open hand is 7♡J♡K◇7◇. What is the probability that you will win the game? How would it change if your hidden card is K♠?

6. You are playing five-card stud poker without bring-in and only one opponent is left. You show 5◇3♠Q♠5♡ and your hidden card is 5♠. Your opponent's open hand is 7♡K◇7◇K♣. What is the probability that you will win the game?

7. In a Texas Hold'em game with only two players, your hand is 8♣J♠, your opponent's hand is 8◇10◇ and the hand shown on the table before the river is 2◇K♠8♠Q♡. What is the probability that you win the hand once the river is turned? What is the probability that you will tie?

8. What would be the answer to the previous problem if you only knew one of your opponent's cards; in particular, what would it be if you only knew that his hand includes 8◇? *Note:* This can be a bit laborious.

9. In a Texas Hold'em game with only two players, your hand is 4◇6♠, your opponent's hand is 10♠10♡ and the hand shown on the table before the river is 7◇10♣8♡5◇. What is the probability that you win the hand? What is the probability that you will tie?

10. In a Texas Hold'em game with only two players, your hand is 10◊J◊, your opponent's hand is 9♠K♣ and the hand shown on the table before the river is 9◊9♡8♣K◊. What is the probability that you win the hand? What is the probability that you will tie?

11. In a Texas Hold'em game with only two players, your hand is 6♣A♡, your opponent's hand is 10♠A◊ and the hand shown on the table before the river is 9◊7♡8♣Q◊. What is the probability that you win the hand? What is the probability that you will tie?

12. In a Texas Hold'em game with only two players, your hand is 10◊10♠, your opponent's hand is unknown and the hand shown on the table K◊K♡2♣K♠. What is the probability that you win the hand? This will be a long calculation; it will be very good if you can start by outlining your solution and then add details as much as possible.

13. [R] Build a simulation to corroborate the calculation of the probability of a flush. How large do you need to make your simulation if you want to get accurate estimates of this probability?

14. [R] The strategy for hand substitutions in draw poker is not always obvious. A common conundrum is the following: Assume that your hand consists of a pair of 2, an A, and two more cards that are not 2, A, or form a pair. Should you keep the pair of 2 and swap three cards, or should you keep the A and swap four cards? Write a simulation in R that can help you decide which option leads to a higher hand.

15. [R] Construct a simulation to estimate the probability that you win a game of Hold'em if you hold K♣2♡ and the five community cards are A◊6♡3♣K♠10♠.

10

Strategic Zero-Sum Games with Perfect Information

So far, we have focused most of our attention on random games where the player is pitted against a non-intelligent opponent. We now turn our attention to *strategic games*, in which two rational opponents attempt to outsmart each other. In this type of situation, players still attempt to maximize their respective utilities, but they now need to cope with their opponent's ability to anticipate their actions and act accordingly. Because of this, the process of choosing the optimal action involves predicting what our opponent will prefer when faced with his own options, and our own optimal strategies need to be devised by conditioning on what we expect our opponent to rationally prefer.

To emphasize the difference, compare the games of blackjack and poker. While playing blackjack, we designed our basic strategy knowing that the dealer will always stay with 17 or more and hit with 16 or less. That made it reasonable for us to stay with relatively low numbers (say 14) as long as the dealer's face-up card suggested that he had a good chance of going bust (e.g., if the dealer shows a 6). This is so because the house always plays the same strategy, and the dealer will hit with 16 even if our current hand is only 14. This is different from the situation in poker, where we need to account for the fact that the other players might be bluffing when they raise their bets.

We start by considering the simplest possible strategic game, which includes two intelligent opponents who try to outwit each other in a game, where the winnings of one player translate into losses for the other. These types of games are called *strategic zero-sum games* because one player's profit equals the loss of its opponents.

10.1 Games with Dominant Strategies

Consider the following strategic "game" involving two companies that sell mineral water. We will call the companies Pevier and Errian. Each company has a fixed cost of $5000 per period, regardless of whether they sell anything or not. The two companies are competing for the same market, and each firm

Probability, Decisions and Games: A Gentle Introduction using R, First Edition. Abel Rodríguez and Bruno Mendes.
© 2018 John Wiley & Sons, Inc. Published 2018 by John Wiley & Sons, Inc.
Companion website: www.wiley.com/go/Rodriguez/Probability_Decisions_and_Games

must choose a *high price* ($2 per bottle) or a *low price* ($1 per bottle). The rules of the game are:

- At a price of $2, a total of 5000 bottles can be sold for a total revenue of $10,000.
- At a price of $1, a total of 10,000 bottles can be sold for a total revenue of $10,000.
- If both companies charge the same price, they split the sales evenly between them.
- If one company charges a higher price, the company with the lower price sells the whole amount, and the company with the higher price sells nothing.
- Both companies aim at maximizing their profit, which is simply the revenue from sales minus the $5000 in fixed costs.

Under these rules, there are four situations that could arise:

- If both companies charge the high price ($2), then 5000 bottles are sold at a price of $2 each, for a total revenue of $10,000 that is split evenly between both companies. Therefore, each company has a revenue of $5000 and a net profit of $0 once the fixed costs are subtracted.
- If both companies charge the low price ($1), then 10,000 bottles are sold at a price of $1 each. Again the total revenue is $10,000, which is split evenly between Errian and Pevier. As before, this leads to both companies having a net profit of $0.
- If Pevier charges the high price and Errian the low price, then all the revenue ($10,000) goes to Errian, which makes a net profit of $5000. In turn, this means that Pevier has no revenue and makes a net loss of $5000.
- Under the same logic, if Pevier charges the low price and Errian charges the high price, then Pevier makes $5000 net profit and Errian loses $5000.

Table 10.1 summarizes the utility (in this case, profit) that each company gets under each of the four combinations of strategies. This type of table is called the *normal form* of the game. The first number on each cell represents Errian's net profit for that combination of strategies, while the second number corresponds

Table 10.1 Profits in the game between Pevier and Errian.

		Pevier	
		Low price	High price
Errian	Low price	$(0, 0)$	$(5000, -5000)$
	High price	$(-5000, 5000)$	$(0, 0)$

to Pevier's net profit. Note that, in every box, the sum of the two numbers is the same (zero in this case). Indeed, this game is an example of a *zero-sum game*.

Zero-Sum Games

In zero-sum games, the sum of utilities for all players in the game, for every combination of strategies, is constant (typically, but not necessarily, zero). More informally, in a zero-sum, a player benefits only at the equal expense of its opponents.

Our goal when dealing with strategic games such as this is to predict how each of the player will behave (in our running example, what price they will choose). We call this prediction the *solution* of the game. To construct our solution, we assume that both companies anticipate the actions of their opponent and act accordingly to try to maximize their own profit. To do this, we first look at the problem from Pevier's perspective by exploring the consequences of its choices:

- First, assume that the Errian decides to set the price of its product to \$1 (i.e., we are in the first row of Table 10.1). In this case, Pevier needs to pick between losing \$5000 (if it chooses the high price) or making no profit (if it chooses the low price). Hence, Pevier's best option (called its *best response*) in this case is to set its price to \$1.
- Next, assume that the Errian decides to set the price of its product to \$2 (i.e., we are in the second row of Table 10.1). Therefore, Pevier will either make no profit (if it chooses to price their water at \$2) or it could potentially make \$5000 profit if it decides to set the price to \$1. Again, Pevier's best alternative is to set the price of its water to \$1.

Hence, no matter what Errian decides to do, Pevier's optimal decision is to set a low price (L), so Errian can reasonably expect that Pevier will do exactly that. Because the game is *symmetric* (i.e., the same reasoning applies if we look at the problem from Errian's perspective), we can predict that Errian will also select to price its water at the low price of \$1 (L). In summary, we can be fairly certain that the rational outcome of the game is for both players to select the low price for their product, which we represent as (L, L). This problem is an example of a game where both players have a *dominant strategy*.

Dominant Strategy

A dominant strategy is a strategy that is at least as good as the alternatives in all circumstances and better in some. When a rational player has a dominant strategy, we can be fairly sure that he will play exactly that strategy.

The solution (L, L) given by the dominant strategies we just found has some interesting properties. For example, imagine that the two companies reassess the price of their products every 6 months (i.e., they play the game multiple times) and they used the (L, L) strategy in the previous round. Then, as long as they believe that the other player will stick to the strategy used this round, none of them has an incentive to change the strategy in the next one. That is, unilateral changes in strategy are not beneficial to any of the players, and therefore extremely unlikely. Note that this property is not shared by any of the other three strategies. In addition, each player's strategy is the best possible response to the other's player action (if Errian sets a low price, the best response that Pevier can adopt is also a low price, and vice-versa). We call pairs of strategies that satisfy these two properties *Nash equilibria*, in honor of John Forbes Nash, whose life has been portrayed in the movie *A Beautiful mind*.

Nash Equilibrium

In a two-person game of perfect information, a pair of strategies (one for each player) is called a Nash equilibrium if they are mutual best responses.

We can rationalize Nash equilibria as the consequence of players learning after playing the game repeatedly. For example, assume that Errian and Pevier play their game multiple times and that in the beginning both players adopt the high-price strategy. After acting in this way for a while, one of the players (say, Errian) is likely to wise up and realize that they can make money out of their opponent by changing to a low price strategy in order to minimize loses. But once the high-price holdout (Pevier) realizes that the other player will stick to the low price, they will also turn to a low price strategy. Once both players have decided to charge a low price, there is no reason for them to change their strategies unilaterally. Note, however, that, this interpretation is useful only for games that can be repeated over and over, just as in the case of the frequentist interpretation of probability discussed in Chapter 1.

The strategies we have used so far involve players repeatedly using one of their actions. This type of strategies are called *pure strategies.*

Pure-strategy Nash equilibrium

We say that a Nash equilibrium involves pure strategies if, in equilibrium, each player always takes the same action.

Note that the process we used to obtain our solution to the game makes a few assumptions about the players. First, we are assuming that all players are rational (i.e., they maximize some utility function). Second, we assume that all players know that the other players are rational, follow the same rules, and know what the utility function of other players are (i.e., rationality and utility functions are common knowledge). Finally, we are assuming that players act simultaneously without the knowledge of the other player's choice (in the Perrier vs Errian example this last assumption did not make a difference, but in the future it might). Unless noted otherwise, we will retain these assumptions in the remainder of this book.

10.2 Solving Games with Dominant and Dominated Strategies

Let's now consider another example related to politics. Two presidential candidates (call them Ling and Matt) are engaged in a debate, and they need to decide where they stand on two conflicting issues (e.g., whether to raise income taxes or not) or whether they will dodge the issue. Assume that, after extensive polls, there is agreement among political analysts on the percentage of the vote each candidate will receive for each combination of positions. Table 10.2 presents these percentages.

As stated earlier, the first number in each cell represents Ling's percentage of the vote, and the second represents Matt's percentage. Even though the sum of the entries in each cell is 100% instead of 0, this is still a zero-sum game. Indeed, since the solution of the game depends on the ordering of the preferences but not their exact values, we could subtract 50 from every entry in Table 10.2, making the values in each entry add to 0 without altering the solution.

Let's consider first the game from Matt's perspective and find his best response to each of Ling's actions. If Ling supports an increase in taxes, Matt should dodge the issue, leaving him with 60% of the vote. On the other hand, if Ling decides not to support an increase in taxes, Matt should also dodge the issue, in which case both candidates will tie. Finally, if Ling decides to dodge the issue, Matt should yet again dodge the issue in order to get 60% of the vote. These observations are summarized in Table 10.3.

Let's turn to Ling's best responses. If Matt decides to support an increase in taxes, Ling should not support it, netting her 60% of the vote. If Matt decides not to support the tax increase, then Ling could either not support it or dodge the issue, which would leave her with 55% of the vote. Finally, if Matt dodges the issue, Ling should not support the increase, again leaving both with 50% of the vote each. Again, we summarize these results in Table 10.4.

Table 10.2 Poll results for Matt versus Ling (first scenario).

		Matt		
		Increase	No increase	Dodge issue
Ling	Increase	(45%, 55%)	(50%, 50%)	(40%, 60%)
	No increase	(60%, 40%)	(55%, 45%)	(50%, 50%)
	Dodge issue	(45%, 55%)	(55%, 45%)	(40%, 60%)

Table 10.3 Best responses for Matt (first scenario).

If Ling decides ...	Matt should ...
... to support an increase in taxes,	... dodge the issue.
... not to support an increase in taxes,	... dodge the issue.
... to dodge the discussion,	... dodge the issue.

Table 10.4 Best responses for Ling (first scenario).

If Matt decides ...	Ling should ...
... to support an increase in taxes,	... not support an increase in taxes.
... not to support an increase in taxes,	... not support it or dodge.
... to dodge the discussion,	... not support an increase in taxes.

Once the best responses have been obtained, the analysis of this game is relatively simple. Note that, no matter what Ling does, Matt should always dodge a discussion about taxes. In addition, note that no matter what Matt does, not supporting a tax increase is always optimal for Ling. In other words, these two strategies are dominant. Therefore, it is reasonable to expect that Ling will not support the increase in taxes, and Matt will avoid any discussion on the topic, leaving the electorate evenly split among the candidates. As before, this solution corresponds to a Nash equilibrium because they are mutual best responses, and therefore there is no incentive for the players to unilaterally alter their strategies.

Consider now a slight modification of this political game where the share of the vote for each candidate is instead given in Table 10.5. In this case, if Ling chooses to support an increase in taxes Matt is better off not supporting the increase. On the other hand, if Ling does not support an increase, Matt should dodge the issue, and if Ling dodges the issue, Matt should not support the increase. From Ling's perspective, if Matt supports an increase, Ling

Table 10.5 Poll results for Matt versus Ling (second scenario).

		Matt		
		Increase	No increase	Dodge issue
	Increase	(45%, 55%)	(10%, 90%)	(40%, 60%)
Ling	No increase	(60%, 40%)	(55%, 45%)	(50%, 50%)
	Dodge issue	(45%, 55%)	(10%, 90%)	(40%, 60%)

Table 10.6 Best responses for Matt (second scenario).

If Ling decides ...	Matt should ...
... to support an increase in taxes,	... not support an increase in taxes.
... not to support an increase in taxes,	... dodge the issue.
... to dodge the discussion,	... not support an increase in taxes.

Table 10.7 Best responses for Ling (second scenario).

If Matt decides ...	Ling should ...
... to support an increase in taxes,	... not support an increase in taxes.
... not to support an increase in taxes,	... not support it or dodge.
... to dodge the discussion,	... not support an increase in taxes.

should not support the increase. If Matt does not support the increase, Ling should again not support the increase. Finally, if Matt dodges the issue of tax increases, then Ling should (for a third time) not support the increase. These results are summarized in Tables 10.6 and 10.7.

Unlike our previous example, Matt does not have a dominant strategy. This might suggest that solving the game is much harder. However, this is not the case. Not supporting the increase is a dominant strategy for Ling; consequently, we can be sure that she will adopt it. Once we know that Ling will not support an increase on taxes, our previous discussion suggests that Matt's rational reaction should be to dodge the issue, which again leads to both candidates splitting the vote 50% each. This solution is again a Nash equilibrium.

Let's consider one last set of payoffs, as given in Table 10.8. Tables 10.9 and 10.10 contain the best responses. In this case, none of the players has a dominant strategy, that is, it is not immediately obvious what any given player should do. However, it is clear what Matt *should not do*. Indeed, Table 10.10 suggests that Matt should never dodge the issue, as dodging is never a best response. Similarly, from Table 10.9 note that supporting a tax increase is

Table 10.8 Poll results for Matt versus Ling (third scenario).

			Matt	
		Increase	No increase	Dodge issue
	Increase	(35%, 65%)	(10%, 90%)	(60%, 40%)
Ling	No increase	(45%, 55%)	(55%, 45%)	(50%, 50%)
	Dodge issue	(40%, 60%)	(10%, 90%)	(65%, 35%)

Table 10.9 Best responses for Ling (third scenario).

If Ling decides ...	Matt should ...
... to support an increase in taxes,	... not support an increase in taxes.
... not to support an increase in taxes,	... support an increase in taxes.
... to dodge the discussion,	... not support an increase in taxes.

Table 10.10 Best responses for Matt (third scenario).

If Matt decides ...	Ling should ...
... to support an increase in taxes,	... not support an increase in taxes.
... not to support an increase in taxes,	... not support an increase in taxes.
... to dodge the discussion,	... also dodge the discussion.

never a good idea for Ling. This observation indicates that dodging (in the case of Matt) and supporting the tax increase (in the case of Ling) are *dominated strategies*.

Dominated Strategy

A strategy that is no better than the alternatives in all circumstances, and worse in some, is called a *dominated strategy*. If a player has a dominated strategy, we can be fairly certain that they will never play it.

Finding dominated strategies can help us solve a game by reducing the number of actions that we need to consider. Indeed, since Matt will never dodge the issue and Ling will never support an increase in taxes, we could simply eliminate the corresponding row and column from the table and work with the *reduced game* (see Table 10.11). In this reduced game, we only need to consider Ling's reactions to Matt supporting or not supporting an increase in taxes and Matt's reactions to Ling not supporting the increase or

Table 10.11 Reduced table for poll results for Matt versus Ling.

		Matt	
		Increase	No increase
Ling	No increase	(45%, 55%)	(55%, 45%)
	Dodge issue	(40%, 60%)	(10%, 90%)

dodging the issue. Hence, it is easy to see that no matter which rational option Matt chooses, Ling's optimal response is not to support an increase in taxes. In other words, even though the original game did not have any dominant strategy, once the dominated strategies are eliminated, not supporting a tax increase becomes a dominant strategy for Ling. The final solution of the game is obtained by noting that Matt's best response to Ling's dominant strategy of not supporting the tax is for Matt to support it, which will lead to 45% of the electorate to support Ling and 55% to support Matt.

10.3   General Solutions for Two Person Zero-Sum Games

When dominant or dominated strategies are present, the solution of a game can often be obtained by applying the two insights we discussed earlier:

- If a strategy is dominant for one player, we can be sure that she will use it, and therefore we only need to look at the best response of the other player to the dominant strategy.
- If a strategy is dominated for one player, we can simply remove the corresponding column or row of the matrix and work with the reduced game.

However, not all games have dominant or dominated strategies, so these tools are not always enough to solve a non-zero sum game. A more general approach to solving games uses the fact that a Nash equilibrium corresponds to the pair of strategies that are mutual best responses. For example, consider the two-person game whose outcomes are presented in Table 10.12. The best responses for each of the two players are summarized in Tables 10.13 and 10.14.

It should be clear from the tables of best responses that there are no dominant or dominated strategies for any of the players. However, note that the pair (D, A) is made of mutual best responses (A is the best response to D and D is the best response to A), and that this is the only pair with this characteristic. Hence the pair (D, A) is the unique pure-strategy Nash equilibrium for this game.

Table 10.12 A game without dominant or dominated strategies.

		Player 2		
		A	B	C
Player 1	D	$(3, -3)$	$(4, -4)$	$(5, -5)$
	E	$(2, -2)$	$(1, -1)$	$(-6, 6)$
	F	$(-1, 1)$	$(5, -5)$	$(-2, 2)$

Table 10.13 Best responses for Player 1 in our game without dominant or dominated strategies.

If Player 2 chooses ...	Player 1 should choose ...
... A,	... D.
... B,	... F.
... C,	... E.

Table 10.14 Best responses for Player 2 in our game without dominant or dominated strategies.

If Player 1 chooses ...	Player 2 should choose ...
... D,	... A.
... E,	... B.
... F,	... C.

Table 10.15 Example of a game with multiple equilibria.

		Player 2		
		A	B	C
Player 1	A	$(0,0)$	$(1,-1)$	$(0,0)$
	B	$(-1,1)$	$(0,0)$	$(-1,1)$
	C	$(0,0)$	$(1,-1)$	$(0,0)$

It is also important to note that Nash equilibria might not be unique. For example, consider the game represented in Table 10.15, which has four equilibria: (A,A), (A,C), (C,A), and (C,C). In the case of zero-sum games, all the equilibria must have the same payoffs, so players will be indifferent among them (but for more general games such as the ones discussed in Chapter 12, the payoffs might be different).

10.4 Exercises

1. A recent New York Times article contained the following statement: "The answer is simple because the zero-sum game nature of our politics demands that one party represent progress and the other the status quo." Explain what the phrase "zero-sum game" means in this context.

2. The following table corresponds to the payoff of a zero-sum game to Player A, when A plays the strategy in the row and B corresponds to the strategy in the column.

		Player B		
		L	M	H
	D	19	0	1
Player A	F	11	9	3
	U	23	7	−3

(a) Is there any dominated strategy for either player?
(b) Is there any dominant strategy for either player?
(c) What is an equilibrium strategy for this game?
(d) What is the payoff for the game?

3. The following table corresponds to the payoff of a zero-sum game to player Liza, when Liza plays the strategy in the row and Jose choices correspond to the strategies in the columns.

		Jose		
		1	2	3
	1	−2	1	1
Liza	2	−3	0	2
	3	−4	−6	4

(a) Is there any dominated strategy for either player?
(b) Is there any dominant strategy for either player?
(c) What is an equilibrium strategy for this game?
(d) What is the payoff for the game?

4. The following table corresponds to the payoff of a zero-sum game to player L. Gaga , when L. Gaga plays the strategy in the row, and the strategies in the columns correspond to choices available to player Shakira.

		Shakira		
		x	y	z
	a	2	1	3
L. Gaga	b	−1	1	2
	c	−1	0	1

(a) Is there any dominated strategy for either player?
(b) Is there any dominant strategy for either player?
(c) What is an equilibrium strategy for this game?
(d) What is the payoff for the game?

5. Show that the game presented in Table 10.15 indeed has the four Nash equilibria (A,A), (A,C), (C,A), and (C,C).

6. In a simplified, single-move sword duel, each player has four different moves: two attacking moves (A1 and A2) and two defensive moves (D1 and D2). Attacking move A1 is very effective against attacking move A2 and defensive move D2 (it leads to a gain of 4 and 3 points, respectively, to the player who uses it). Defensive move D1 is very effective against attacking move A1 (leads to a gain of 2 points), it's a poor move against A2 (−1 point) and is marginally better than defensive move D2 (1 victory point). Finally, defending move *D*2 does very badly against attacking move A2 (−3 points). When the two players choose the same move, the result is a draw (0 points each), and whatever one player wins/looses the other player looses/wins (respectively). Is this a two-person zero-sum game? Is there an equilibrium point for this game?

7. *Even or odd?* is a children's two-person game in Portugal. Players take turns saying "even" (or "odd"); on the count of three, players show a number with their hand; one wins if the sum of both numbers is even and one said "even", or if the sum is odd and you said odd. A draw occurs if both players guess right, or if both guess wrong. Is this a zero-sum game? Why? Set-up up the game in normal form and see if it has a pure strategy.

8. Two players are bargaining over how to split 1. Both players simultaneously name shares they would like to have, *x* and *y*, where $0 \le x, y \le 1$. If the sum of the shares is less than one, each one receives the shares they named. On the other hand, if the sum is greater than 1 , then both players receive zero. Show that a 50%–50% split is a pure-strategy Nash equilibrium for this problem. Is this equilibrium unique? To answer this question, assume that the utility function of the players is strictly monetary (i.e., they do not derive any utility from "screwing" their opponents). *Hint:* Recall the definition of a Nash equilibrium as a pair of actions that are mutual best responses. Is a 30%–70% split an equilibrium point? Is a 1%–99% split an equilibrium point?

11

Rock–Paper–Scissors: Mixed Strategies in Zero-Sum Games

Not all two-person, zero-sum games admit a pure-strategy equilibrium (i.e., a solution that results in the players always using the same strategies over and over again). A well-known example is the game of *rock–paper–scissors*, also known as *roshambo*. Rock–paper–scissors is a hand game between two players that simultaneously select among three gestures (corresponding to *rock*, *paper*, or *scissors*). The objective of the game is to select a gesture that defeats that of the opponent, and the winner of each round makes $1 out of the other player. The game is resolved as follows:

- Rock breaks scissors, so rock defeats scissors.
- Scissors cut paper, therefore, scissors defeats paper.
- Paper covers rock, hence, paper defeats rock.
- If both players select the same gesture, the game results in a tie.

The payoffs of the game for our two players Jiahao and Antonio are reported in Table 11.1. Since the outcomes for the two players add to the same quantity in all cell, this is clearly a zero-sum game. As we have done before, our next step in solving the game is to come up with optimal responses from each player's point of view (which are probably already obvious to you). Table 11.2 shows the best responses for Jiahao. Because of the symmetry of the game, the same table also applies to Antonio.

Note that no combination of strategies leads to a pure-strategy equilibrium: once Jiahao knows that Antonio will play (say) rock for sure, he has a clear incentive to continuously play paper. But once Jiahao realizes that Antonio will play paper for sure, he has an incentive to start playing scissors. In turn, this will lead to Antonio selecting to play rock. Consequently, players who repeatedly play this game will tend to continuously cycle through the different gestures. This is unlike the examples we discussed before where, once the equilibrium strategy has been attained, there is no incentive for the players to deviate unilaterally.

If you have ever played rock–paper–scissor before, you have probably figured out that constantly changing your gesture in a more or less unpredictable

Probability, Decisions and Games: A Gentle Introduction using R, First Edition. Abel Rodríguez and Bruno Mendes.
© 2018 John Wiley & Sons, Inc. Published 2018 by John Wiley & Sons, Inc.
Companion website: www.wiley.com/go/Rodriguez/Probability_Decisions_and_Games

Table 11.1 Player's profit in rock–paper–scissors.

		Antonio		
		Rock	Paper	Scissors
	Rock	$(0,0)$	$(-1,1)$	$(1,-1)$
Jiahao	Paper	$(1,-1)$	$(0,0)$	$(-1,1)$
	Scissors	$(-1,1)$	$(1,-1)$	$(0,0)$

Table 11.2 Best responses for Jiahao in the game of rock–paper–scissors.

If Antonio plays ...	Jiahao should play ...
... rock,	... paper.
... paper,	... scissors.
... scissors,	... rock.

manner is a better strategy than always selecting the same gesture. Indeed, it turns out that the optimal strategy for this game corresponds to randomly selecting a strategy among the three available to them. How often should they play each strategy in rock–paper–scissor? Because the profit from playing the game is the same ($1) under all strategies, the (correct!) intuition is that players should alternate among strategies so that you spend about 1/3 of their time playing each one of them. Game strategies that involve such randomly chosen actions each time the game is played are called *mixed strategies*. This is in contrast to the pure strategies we studied in the previous chapter, which involve always using the same action.

> **Mixed-Strategy Nash Equilibria**
> We say that a Nash equilibria involves mixed strategies if, in equilibrium, each player randomizes their actions each time they play according to a given probability distribution over their options.

In real-life games, mixed-strategy equilibria sometimes fail to materialize as the long-term outcome of repeated games because humans are very bad at avoiding patterns of behavior, but they are nonetheless the optimal way to proceed.

11.1 Finding Mixed-Strategy Equilibria

A general approach for deriving mixed-strategy equilibria in zero-sum games is to find a set of probabilities for the actions available to each player such that a player is indifferent to which strategy she selects if the opponent randomizes according to their probabilities, and vice-versa. Indeed, if both players are willing to randomize, then the expected utility associated with each alternative should be the same (otherwise, the players would not randomize but would always choose the option that maximizes their utility).

Let's illustrate this principle using the rock–paper–scissors example. Let q_r be the probability that Antonio chooses rock on a given round, q_p the probability that he chooses paper, and q_s the probability that he chooses scissors. Note that, since these are the only three options available, these probabilities need to satisfy $q_r + q_p + q_s = 1$. The expected value of each strategy for Jiahao is then given in Table 11.3.

Now, remember that no strategy by itself is the best one among all strategies; this means that the rational player will have no particular preference among any of their strategies. This in turn means that the expected values for all strategies need to be equal to each other. For example, we could make the expected value of the game for rock and paper, as well as that for rock and scissor, equal. This leads to

$$\text{From the first and second lines,} \quad q_s - q_p = q_r - q_s, \tag{11.1}$$

$$\text{From the first and third lines,} \quad q_s - q_p = q_p - q_r, \tag{11.2}$$

which, together with the fact that all probabilities need to add up to one,

$$q_s + q_p + q_r = 1 \tag{11.3}$$

gives us a system of three equations with three unknowns (with the unknowns corresponding to the probabilities Antonio will choose each strategy). To solve the system of equations, first add (11.1) and (11.2) together to get

$$2q_s - 2q_p = q_p - q_s \quad \Leftrightarrow \quad 3q_p = 3q_s \quad \Leftrightarrow \quad q_p = q_s. \tag{11.4}$$

Inserting this result back into (11.1), we also get

$$q_s - q_s = q_r - q_s \quad \Leftrightarrow \quad q_s = q_r.$$

Table 11.3 Utility associated with different actions that Jiahao can take if he assumes that Antonio selects rock with probability q_r, paper with probability q_p and scissors with probability q_s.

If Jiahao plays ...	The expected value of the game for Jiahao is ...
... rock,	... $0 \times q_r + (-1) \times q_p + 1 \times q_s = q_s - q_p$.
... paper,	... $1 \times q_r + 0 \times q_p + (-1) \times q_s = q_r - q_s$.
... scissors,	... $(-1) \times q_r + 1 \times q_p + 0 \times q_s = q_p - q_r$.

Therefore, we have shown that all three probabilities need to be equal. Since they also need to add up to 1 (recall Equation (11.3)), we get

$$q_r + q_r + q_r = 1 \quad \Leftrightarrow \quad 3q_r = 1 \quad \Leftrightarrow \quad q_r = \frac{1}{3} = q_p = q_s.$$

Because of the symmetry of the game, the same argument applies to the randomization strategy of Jiahao, and the equilibrium point of the game corresponds to each player randomly (and independently) selecting a hand gesture with equal probability among the three options available to them every time they play. The expected value of this game is then

$$E \left(\begin{array}{c|c} \text{Utility in} & \text{Players adopt the} \\ \textit{Rock–paper–} & \text{mixed-strategy} \\ \textit{scissors} & \text{equilibrium} \end{array} \right)$$
$$= 0 \times \frac{1}{3} + (-1) \times \frac{1}{3} + 1 \times \frac{1}{3} = 0.$$

Consequently, if the game is repeated multiple times and Jiahao adopts the optimal mixed strategy, he is bound to at least not lose money in the long run. That is, if the players adopt the Nash equilibrium as their strategy, then none of them will make any money. To corroborate this, the following simulation allows you to compare the optimal strategy implied by the Nash equilibrium (which Jiahao always plays) against other strategies played by Antonio.

```
> n = 50000
> opt = c("P", "R", "S")
> player1strat = c(1/3, 1/3, 1/3)
> player2strat = c(0.1, 0.8, 0.1)
> outcome = rep(0, n)      # From Jiahao perspective
> for(i in 1:n){
+     play1 = sample(opt,1,replace=T,prob=player1strat)
+     play2 = sample(opt,1,replace=T,prob=player2strat)
+     if(play1=="P"){
+       if(play2=="S"){
+         outcome[i] = "L"
+       }else{
+         if(play2=="R"){
+           outcome[i] = "W"
+         }else{
+           outcome[i] = "T"
+         }
+       }
+     }else{
+       if(play1=="R"){
+         if(play2=="S"){
+           outcome[i] = "L"
+         }else{
```

```
+              if(play2=="P"){
+                  outcome[i] = "W"
+              }else{
+                  outcome[i] = "T"
+              }
+          }
+      }else{
+          if(play2=="R"){
+              outcome[i] = "L"
+          }else{
+              if(play2=="S"){
+                  outcome[i] = "W"
+              }else{
+                  outcome[i] = "T"
+              }
+          }
+      }
+  }
+  }
> profit = (outcome=="W") - (outcome=="L")
> mean(profit)

[1] -0.00064
```

In the simulation above, we assumed that Antonio picks paper 10% of the time, rock 80% of the time, and scissors 10% of the time, but the result is the same no matter what Antonio does: the long-run profit for both Antonio and Jiahao is always zero. In this game, the Nash-equilibrium can be thought of the best defensive strategy: no matter what the other player does, you cannot lose money. The other side of that statement is that, if the other player uses the Nash equilibrium as its strategy, then there is nothing you can do to make more money.

Pure strategies are a special case of mixed strategies where the probability associated with one of the alternatives is equal to 1. By expanding the space of possible strategies to include mixed strategies, we can guarantee that a large class of zero-sum games has a solution given by a Nash equilibrium. This is a consequence of the *minimax theorem*:

Minimax Theorem

For every two-person, zero-sum game with finitely many strategies, there exists a value V and a mixed strategy for each player, such that (1) given the strategy for Player 2, the best payoff possible for Player 1 is V, and (2) given the strategy for Player 1, the best payoff possible for Player 2 is $-V$.

In simpler words, the minimax theorem states that every two-person, zero-sum game with finitely many strategies has at least one solution, which might involve players using mixed strategies.

11.2   Mixed Strategy Equilibria in Sports

Mixed strategy equilibria appear in a number of sports including baseball, football, and soccer. For example, suppose now that you are playing soccer and, in particular, that you will be kicking a penalty. Therefore, you need to decide how you are going to kick (your options are to kick left, center, or right). The goalkeeper also needs to make a similar decision about where to lunge (again, left, center, or right). Table 11.4 presents the utility derived by each player from each combination of strategies (the numbers correspond to the (historical) conditional probabilities that a goal is scored/not scored).

As with rock–paper–scissors, there are no pure-strategy Nash equilibria for this game. If you have ever watched soccer, this should not be surprising, a kicker or a goal keeper who becomes predictable are typically very bad for their teams.

To determine the mixed-strategy equilibrium we proceed as before and first compute the expected value of the game for the kicker when the goalkeeper randomizes their actions so that with probability q_l she will lunge left, with probability q_c she will stay in the center, and with probability q_r she will lunge right. The results are presented in Table 11.5.

Since the expected utilities must be the same, equating the first two expressions leads to

$$0.65 \times q_l + 0.95 \times q_c + 0.95 \times q_r = 0.95 \times q_l + 0 \times q_c + 0.95 \times q_r$$

Table 11.4   Utilities associated with different penalty kick decisions.

		Goal keeper		
		Left	Center	Right
Kicker	Left	(0.65,0.35)	(0.95,0.05)	(0.95,0.05)
	Center	(0.95,0.05)	(0,1)	(0.95,0.05)
	Right	(0.95,0.05)	(0.95,0.05)	(0.65,0.35)

Table 11.5   Utility associated with different actions taken by the kicker if he assumes that goal keeper selects left with probability q_l, center with probability q_c, and right with probability q_r.

If Kicker kicks ...	the expected value of the game for the kicker is ...
... left,	... $0.65 \times q_l + 0.95 \times q_c + 0.95 \times q_r$.
... center,	... $0.95 \times q_l + 0 \times q_c + 0.95 \times q_r$.
... right,	... $0.95 \times q_l + 0.95 \times q_c + 0.65 \times q_r$.

$$\Leftrightarrow \qquad 0.95 \times q_c = 0.30 \times q_l$$

$$\Leftrightarrow \qquad q_c = \tfrac{0.30}{0.95} \times q_l.$$

Similarly by equating the second and third expressions, we have

$$0.95 \times q_l + 0 \times q_c + 1 \times 0.95 q_r = 0.95 \times q_l + 0.95 \times q_c + 1 \times 0.65 q_r$$

$$\Leftrightarrow \qquad 0.95 \times q_c = 0.30 \times q_r$$

$$\Leftrightarrow \qquad q_c = \frac{0.30}{0.95} \times q_r.$$

Note that one consequence of these two equations is that $q_r = q_l$. This makes intuitive sense; if we look at the payoff table, we realize that the left and right choices are interchangeable. Now, using the fact that $q_r + q_c + q_l = 1$, we have

$$\frac{0.95}{0.30} \times q_c + q_c + \frac{0.95}{0.30} \times q_c = 1$$

$$\Leftrightarrow \qquad \frac{0.95 + 0.30 + 0.95}{0.30} q_c = 1$$

$$\Leftrightarrow \qquad q_c = \frac{0.30}{2.2} \approx 0.1363,$$

and from there we get

$$q_l = q_r = \frac{0.95}{0.30} \times \frac{0.30}{2.20} = \frac{0.95}{2.20} \approx 0.4318.$$

Therefore, the equilibrium strategy for the goalkeeper is to lunge to the left about 43% of the time, to the right around 43% of the time, and to stay in the center about 14% of the time. If we now look at the kicker's optimal strategy, we discover that it is identical (he should kick to the right 43% of the time, to the left 43% of the time, and to the center 14% of the time). The expected value of the game for the kicker, if both players stick to this strategy, is obtained by inserting the optimum probabilities found above in any of the expressions for the expected value of the game in Table 11.5 (recall that, by definition, they have to be the same)

$$0.4318 \times 0.95 + 0.4318 \times 0.95 \approx 0.82045.$$

This number can be interpreted as the (marginal) probability of scoring a goal if the players follow the optimal strategy.

11.3 Bluffing as a Strategic Game with a Mixed-Strategy Equilibrium

We turn our attention now to a very simplified version of "poker" in which you and your opponent Alya each place a \$5 bet on the table and then secretly toss a coin with a 0 on one side and 1 on the other. You play first, and you can decide to either *pass* (P) or *bet* (B) an additional \$3. If you pass, the numbers tossed

Figure 11.1 Graphical representation of decisions in a simplified version of poker.

by you and Alya are compared; the largest number takes the pot ($10). If both numbers are the same, each player gets their $5 back (a tie). On the other hand, if you bet the extra $3, Alya might decide to *see* (*S*) or *fold* (*F*). If Alya folds, you get the pot ($13, of which $8 are yours and $5 are Alya's) irrespective of the numbers tossed. If Alya decides to see she must add $3 to the pot (for a total of $16). Again the numbers are compared; the larger number takes the $16 and if the numbers are equal, each gets their money back. Figure 11.1 shows a decision tree with the sequence of decisions associated with the game.

To analyze this game, we consider both rounds of bets simultaneously and write down all strategies available to each player. Each of these strategies will describe what action the player takes based on what the coin shows. For example, you could decide to pass no matter whether you have a 0 or a 1 (call this strategy *pass-pass*, or *PP*). Alternatively, you could pass whenever you have a 0 and bet if you have a 1 (call this *pass-bet*, or *PB*), you can bet if you have a 0 and pass if you have a 1 (call this *bet-pass*, or *BP*), or you could always bet no matter what the outcome of the toss is for you (call this strategy *bet-bet*, or *BB*). Intuitively, some of these strategies are very bad (e.g., playing *BP* is clearly a bad idea); this will be confirmed by our analysis below. Similarly, Alya also has four strategies, *FF* (fold no matter what her coin shows), *SS* (always see, no matter what her coin shows), *FS* (fold is she has a 0 and see if she has a 1), and *SF* (see if she has a 0, and fold if she has a 1).

The outcome of the game for every pair of strategies is summarized in Table 11.6. The numbers in the table correspond to the expected profit for each player from each particular combination of strategies. For example, if you

Table 11.6 Expected profits in the simplified poker.

		Alya			
		FF	*SS*	*SF*	*FS*
You	*PP*	$(0,0)$	$(0,0)$	$(0,0)$	$(0,0)$
	BB	$(5,-5)$	$(0,0)$	$(4.5,-4.5)$	$(0.5,-0.5)$
	PB	$(1.25,-1.25)$	$(0.75,-0.75)$	$(2,-2)$	$(0,0)$
	BP	$(3.75,-3.75)$	$(-0.75,0.75)$	$(2.5,-2.5)$	$(0.5,-0,5)$

decide to play *PP*, you will always pass and the outcome will depend only on the outcomes of the coin tosses no matter what strategy Alya chooses (as she never gets to play). Therefore, all the cells in the first row of the table are equal to zero,

$$
E\begin{pmatrix} \text{Profit for you from playing} \\ PP \text{ no matter what strategy} \\ \text{Alya follows} \end{pmatrix} = \underbrace{5}_{\substack{\text{Your profit if} \\ \text{you get a 1 and} \\ \text{Alya} \\ \text{gets a 0}}} \times \underbrace{\frac{1}{4}}_{\substack{\text{Probability that} \\ \text{you get a 1 and} \\ \text{Alya gets a 0}}}
$$

$$
+ \underbrace{(-5)}_{\substack{\text{Your profit if} \\ \text{you get a 0 and} \\ \text{Alya} \\ \text{gets a 1}}} \times \underbrace{\frac{1}{4}}_{\substack{\text{Probability that} \\ \text{you get a 0 and} \\ \text{Alya gets a 1}}} + \underbrace{0}_{\substack{\text{Your profit if} \\ \text{both get a 1} \\ \text{or both get a 0}}} \times \underbrace{\frac{2}{4}}_{\substack{\text{Probability that} \\ \text{both get a 1} \\ \text{or both get a 0}}} = 0.
$$

Similarly, the expected value for you when you choose strategy *BB* and Alya chooses *FF*,

$$
E\begin{pmatrix} \text{Profit for } BB \text{ and} \\ \text{Alya chooses } FF \end{pmatrix} = \underbrace{5}_{\substack{\text{Your profit if} \\ \text{your get a 1 and} \\ \text{Alya} \\ \text{gets a 0}}} \times \underbrace{\frac{1}{4}}_{\substack{\text{Probability that} \\ \text{you get a 1 and} \\ \text{Alya gets a 0}}}
$$

$$
+ \underbrace{5}_{\substack{\text{Your profit if} \\ \text{you get a 0 and} \\ \text{Alya} \\ \text{gets a 1}}} \times \underbrace{\frac{1}{4}}_{\substack{\text{Probability that} \\ \text{you get a 0 and} \\ \text{Alya gets a 1}}} + \underbrace{5}_{\substack{\text{Your profit if} \\ \text{both get a 1}}} \times \underbrace{\frac{1}{4}}_{\substack{\text{Probability that} \\ \text{both get a 1}}}
$$

$$
+ \underbrace{5}_{\substack{\text{Your profit if} \\ \text{both get a 0}}} \times \underbrace{\frac{1}{4}}_{\substack{\text{Probability that} \\ \text{both get a 0}}} = 5,
$$

while the expected value for you when you choose strategy *BB* and Alya chooses *SS*,

$$
E\begin{pmatrix} \text{Profit for } BB \text{ and} \\ \text{Alya chooses } SS \end{pmatrix} = \underbrace{8}_{\substack{\text{Your profit if} \\ \text{your get a 1 and} \\ \text{Alya} \\ \text{gets a 0}}} \times \underbrace{\frac{1}{4}}_{\substack{\text{Probability that} \\ \text{you get a 1 and} \\ \text{Alya gets a 0}}}
$$

$$+ \underbrace{0}_{\substack{\text{Your profit if} \\ \text{you get a 0 and} \\ \text{Alya} \\ \text{gets a 1}}} \times \underbrace{\frac{1}{4}}_{\substack{\text{Probability that} \\ \text{you get a 0 and} \\ \text{Alya gets a 1}}} + \underbrace{0}_{\substack{\text{Your profit if} \\ \text{both get a 1}}} \times \underbrace{\frac{1}{4}}_{\substack{\text{Probability that} \\ \text{both get a 1}}}$$

$$+ \underbrace{(-8)}_{\substack{\text{Your profit if} \\ \text{both get a 0}}} \times \underbrace{\frac{1}{4}}_{\substack{\text{Probability that} \\ \text{both get a 0}}} = 5.$$

The remaining entries of Table 11.6 can be computed in a similar way. Once these calculations have been completed, we can proceed to find best responses for each player actions (see Tables 11.7 and 11.8). From these tables, it is clear that *PP* and *BP* are dominated strategies. This is intuitively clear: if you are the first player to play, passing no matter what your hand looks like or always betting in the first round and always folding on the second are very bad ideas. Similarly, for Alya, *FF*, and *SF* are dominated strategies. Again this makes sense: always folding is a bad idea for Alya, as is betting when she has a 0 and folding when she has a 1 (she has some chance of winning if her coin shows a 1, while the best she can do if she sees a 0 is to tie).

If we eliminate the strategies that are dominated, we end up with a reduced payoff table (see Table 11.9). With this reduced table, it is clear that there is no pure-strategy equilibrium for the game. To find a mixed-strategy equilibrium,

Table 11.7 Best responses for you in the simplified poker game.

If Alya plays …	you should play …
… *FF*,	… *BB*.
… *SS*,	… *PB*.
… *SF*,	… *BB*.
… *FS*,	… *BB* or *BP* (you are indifferent).

Table 11.8 Best responses for Alya in the simplified poker game.

If you play …	Alya should play …
… *PP*,	… *FF, SS, SF, FS* (she is indifferent, she never gets to play).
… *BB*,	… *SS*.
… *PB*,	… *FS*.
… *BP*,	… *SS*.

Table 11.9 Expected profits in the simplified poker game after eliminating dominated strategies.

		Alya	
		SS	FS
You	BB	$(0, 0)$	$(0.5, -0.5)$
	PB	$(0.75, -0.75)$	$(0, 0)$

Table 11.10 Expected profits associated with different actions you take if you assume that Alya will select SS with probability q_{SS} and FS with probability q_{FS}.

If you play ...	the expected value of the game for you is ...
... BB,	... $0 \times q_{SS} + 0.50 \times q_{FS} = 0.50 \times q_{FS}$.
... PB,	... $0.75 \times q_{SS} + 0 \times q_{FS} = 0.75 \times q_{SS}$.

we apply the same ideas we have used before (but on the reduced table, since the dominated strategies have been discarded). First, let q_{SS} be the probability that Alya picks the SS strategy, and q_{FS} be the probability that Alya picks the FS strategy. The expected profits for each of your actions if Alya randomizes among SS and FS according to q_{SS} and q_{FS} can be seen in Table 11.10.

Since the expected utility from both actions must be the same in equilibrium, we have

$$0.75 \times q_{SS} = 0.50 \times q_{FS} \quad \Leftrightarrow \quad q_{SS} = \frac{0.50}{0.75}q_{FS} = \frac{2}{3}q_{FS}.$$

Finally, using the fact that $q_{SS} + q_{SF} = 1$, we have

$$\frac{2}{3}q_{FS} + q_{FS} = 1 \quad \Leftrightarrow \quad \frac{2+3}{3}q_{FS} = 1 \quad \Leftrightarrow \quad q_{FS} = \frac{3}{5} = 0.6$$

and $q_{SS} = \frac{2}{5} = 0.4$. In other words, Alya should always see if she has a 1 and should fold 60% of the time if she has a 0.

A similar calculation for your randomization probability shows that you should play BB 60% of the times and PB the other 40% of the times. That is, you should always bet if you have a 1, and bluff 60% of the times in which you got a 0. Substituting these numbers back into the formulas for the expected values we have that your expected profit from this game is

$$E(\text{Your profit}) = 0.75 \times q_{PB} = \frac{3}{4} \times \frac{2}{5} = \frac{3}{10} = 0.3.$$

That is, you win at least 30 cents on average if you play according to your optimal mixed strategy, and Alya looses at most 30 cents on average if she plays using their optimal strategy. None of you can do better than that. It is worthwhile to observe that your payoff is positive in this game because you are able to play first. The fact that the first player has an advantage when bluffing is the reason why the role of dealer in Poker rotates around all players. In addition, note that if the players were to choose their strategies simultaneously then both players would have zero expected value (the game would be very similar to matching pennies, see Exercise 1).

The following simulation can help you corroborate that players have no incentive to unilaterally deviate from the Nash equilibrium. In particular, as long as you stick to your optimal strategy (always betting if you get a 1, and betting 60% of the time if you get a 0), there is nothing that Alya can do to improve her outcome.

```
> n = 1000000
> coinspc = c(0, 1)
> raisspc = c(TRUE,FALSE)
> youstr = c(0.4, 0.6)
> oppstr = c(0.9, 0.1)
> profit = rep(0, n)     # From your perspective
> for(i in 1:n){
+     yourpot = 5
+     opponentpot = 5
+     coin1 = sample(coinspc, 1)
+     coin2 = sample(coinspc, 1)
+     if(coin1==1){   # You bet
+         bet = TRUE
+         yourpot = yourpot + 3
+     }else{   # You bluff
+         bet = sample(raisspc, 1, replace=TRUE, prob=youstr)
+         yourpot = yourpot + 3*bet
+     }
+     if(bet){
+         if(coin2==0){  # Opponent sees sometimes
+             see = sample(raisspc, 1, replace=TRUE, prob=youstr)
+             if(see==TRUE){
+                 opponentpot = opponentpot + 3
+                 compare = TRUE
+             }else{
+                 compare = FALSE
+             }
+         }else{   # Opponent sees for sure
+             opponentpot = opponentpot + 3
+             compare = TRUE
+         }
+     }else{
+         compare = TRUE
+     }
```

```
+    if(compare){
+      if(coin1>coin2){
+        profit[i] = opponentpot
+      }else{
+        if(coin1<coin2){
+          profit[i] = -yourpot
+        }else{
+          profit[i] = 0
+        }
+      }
+    }else{
+      profit[i] = opponentpot
+    }
+  }
> mean(profit)

[1]  0.309276
```

11.4 Exercises

1. Matching pennies is a game played between two players, Player A and Player B. Each player has a penny and must secretly turn the penny to heads or tails. The players then reveal their choices simultaneously. If the pennies match (both heads or both tails), Player A receives one dollar from Player B (+1 for A, −1 for B). If the pennies do not match (one heads and one tails), Player B receives one dollar from Player A (−1 for A, +1 for B). Is there any pure-strategy equilibrium for this game? If so, what is the expected payoff? Is there any mixed-strategy equilibrium for this game? If so, what is the expected payoff?

2. In addition to the pure-strategy equilibrium already discussed, the game whose payoff was given in Table 10.12 also admits mixed-strategy equilibria. Find these equilibrium strategies along with the payoff of the game.

3. Combat and strategy-based video games frequently feature cycles in their characters' or units' effectiveness that resemble the pattern of rock–paper–scissors. These often attempt to emulate cycles in real-world combat (such as where cavalry are effective against archers, archers have an edge over spearmen, and spearmen are strongest against cavalry). It is claimed that this kind of strategy makes the game self-balancing. Explain this claim.

4. Recall the sword duel game described in Exercise 6 from Section 10.4. Back then we found that there's no pure-strategy equilibrium. Find the mixed-strategy equilibria and their expected payoff.

5. In the website http://www.samkass.com/theories/RPSSL.html, an extended version of rock–paper–scissors is described (it was made popular in the TV show *The Big Bang Theory*). Setup the table of payoff for this game and find its solution.

6. What about the seven-gesture version of rock–paper–scissors in http://www.umop.com/rps7.htm?

7. Assume that we are playing the simplified "poker" where the blind is $4 (instead of $5) and the raise is another $4 (instead of $3). How should we randomize and what would be expected value of the game for the player that gets to act first? What would be the optimal strategy for the first player if they knew that the second player will always see on a 1 but will only do so 20% of the time on a 2?

8. [R] Write code to simulate the simplified game of poker described in the previous exercise. What happens if the first player to bet deviates from their optimal strategy, but the second player does not?

9. Why is it important in poker for players to alternate at playing blind bets?

10. Change the rock–paper–scissors game so that when scissor matches paper, scissors gets a gain of 2 and paper a loss of −2 and when rock matches scissors, rock gets a gain of 3 and scissors gets −3. What is the Nash equilibrium for this game? What is the expected payoff to the players?

11. [R] Write code to simulate the penalty kick game discussed at the beginning of Section 11.2.

12. Do all calculations required to construct Table 11.6 and check that your results agree with those provided in this book.

12

The Prisoner's Dilemma and Other Strategic Non-zero-sum Games

The analysis of zero-sum games is relatively st'raightforward because in that kind of games trying to maximize the utility of one of the players is equivalent to minimizing the utility of the other (recall that in zero-sum games whatever one player won was always at the opponent's expense). We will now move away from these purely confrontational games and consider games where the interests of the players are at least partially aligned (e.g., they can earn some benefits without necessarily making their opponents worse off). We call these games *non-zero-sum games* because the sum of the outcomes for both players is not necessarily identical under all combinations of strategies. Non-zero-sum games sometimes lead to unexpected conclusions. In fact, we might have to stop thinking about Nash equilibria as providing the *optimal* solution for the game.

12.1   The Prisoner's Dilemma

To motivate non-zero-sum games, consider the famous *prisoner's dilemma*, which appears in pretty much any procedural show on TV. Two men suspected of committing a crime together are arrested and placed in separate interrogation rooms. This game assumes the police can only prosecute for the more serious charge if one or both suspects confess; otherwise, they can only prosecute them for a lesser charge. Each suspect may either confess or remain silent, and each one knows the consequences of their actions. If one suspect confesses but the other does not, the one who confessed turns incriminate evidence against their partner and goes free, while the other goes to jail for 20 years. On the other hand, if both suspects confess, then both of them go to jail for 5 years. Finally, if both suspects remain silent, they both go to jail for a year for a lesser charge. Assuming that each criminal only cares for their own well-being, the payoffs can be summarized in Table 12.1. Note that the sum of the payoffs is not constant (it is −10 if both confess but −2 if both remain silent) and therefore this is not a zero-sum game.

Probability, Decisions and Games: A Gentle Introduction using R, First Edition. Abel Rodríguez and Bruno Mendes.
© 2018 John Wiley & Sons, Inc. Published 2018 by John Wiley & Sons, Inc.
Companion website: www.wiley.com/go/Rodriguez/Probability_Decisions_and_Games

Table 12.1 Payoffs for the prisoner's dilemma.

		Prisoner 2	
		Confess	Remain silent
Prisoner 1	Confess	$(-5, -5)$	$(0, -20)$
	Remain silent	$(-20, 0)$	$(-1, -1)$

Table 12.2 Best responses for Prisoner 2 in the prisoner's dilemma game.

If Prisoner 1...	Prisoner 2 should...
... confesses,	... confess.
... remains silent,	... confess.

From a cursory examination of the table, it would seem like remaining silent is the optimal solution for the game (at least from the point of view of the aggregate number of years that the criminals will spend in jail). However, this solution is not a Nash equilibrium, and it is unlikely that players will adopt such strategy. To see why, let's consider the set of best responses for Prisoner 2, which are presented in Table 12.2. Note that confessing is a dominant strategy for Prisoner 2 (and, by the symmetry of the game, for Prisoner 1 as well). Hence, the Nash equilibrium corresponds to both Prisoners confessing and getting 5 years of jail each.

This can seem somewhat contradictory at first sight. The outcome that involves both prisoners confessing is an equilibrium, as no player has a unilateral incentive to change their behavior if they know that the other prisoner will confess. However, one can't help but notice how this is clearly a stupid strategy for both prisoners because both would be better off if they could coordinate their actions so that both remained silent. This type of apparent contradiction (where Nash equilibria are not necessarily *good* solutions to the game) do not happen in zero-sum games, but are very common in non-zero-sum games. They arise because in non-zero-sum games details such as the order of play and the ability of the players to communicate, make binding agreements or set side payments can have a important effect on the outcome.

12.2 The Impact of Communication and Agreements

Consider now the non-zero-sum game between Anil and Anastasiya presented in Table 12.3. We assume that no communication, agreements, or wealth

Table 12.3 Communication game in normal form.

		Anastasiya	
		Strategy A	Strategy B
Anil	Strategy 1	$(0, 0)$	$(10, 5)$
	Strategy 2	$(5, 10)$	$(0, 0)$

Table 12.4 Best responses for Anastasiya in the communication game.

If Anastasiya...	Anil should...
... plays A,	... play 2.
... plays B,	... play 1.

Table 12.5 Best responses for Anil in the communication game.

If Anil...	Anastasiya should...
... plays 1,	... play B.
... plays 2,	... play A.

transfers are allowed between the players; these are assumptions that we will make throughout this book unless noted otherwise.

As with all other games, let's consider the set of best responses for each player (see Tables 12.4 and 12.5). The best responses are obtained by noting that, if Anil decides to use strategy 1, then Anastasiya will choose strategy B (which pays 5) over strategy A (which pays 0), while if Anil decides to use strategy 2, then Anastasiya will choose strategy A (which pays 10) over strategy B (which pays 0). Similarly, if Anastasiya decides to use strategy A, then Anil will choose strategy 2 (which pays 5) over strategy 1 (which pays 0), while if Anastasiya decides to use strategy B, then Anil will choose strategy 1 (which pays 10) over strategy 2 (which pays 0).

From Tables 12.4 and 12.5, it is clear that both $(2, A)$ and $(1, B)$ are pure-strategy equilibria for this game. Indeed, A is the best response to 2 and vice-versa, while B is the best response to 1 and vice-versa. In addition, the game admits a third, mixed-strategy equilibrium. To find this additional equilibrium point, let p be the probability that Anastasiya plays A (so that the probability that she plays B is $1 - p$). The expected payoffs for Anil are shown in Table 12.6.

Table 12.6 Expected utility for Anil in the communication game.

If Anil plays ...	the expected value of the game for Anil is ...
... strategy 1,	... $0 \times p + 10 \times (1 - p) = 10 - 10p$.
... strategy 2,	... $5 \times p + 0 \times (1 - p) = 5p$.

Thus, for Anil to be indifferent among his strategies, we need

$$5p = 10 - 10p \quad \Leftrightarrow \quad 15p = 10 \quad \Leftrightarrow \quad p = \frac{10}{15} = \frac{2}{3}$$

and the expected payoff for Anil is $5p = \frac{10}{3}$. Exactly the same argument applies to Anastasiya. Note that this is really a Nash equilibrium for the game. If Anil plays 1 with probability 2/3, there is nothing that Anastasiya can do to improve her expected utility over the one she would get by playing A 2/3 of the time, and vice-versa.

Note that, unlike the pure-strategy equilibria, the mixed-strategy equilibrium is fair (in the sense that the payoff for both players is the same, $\frac{10}{3}$). However, the expected payoff of $\frac{10}{3} \approx 3.333$ for each player is still well below the payoffs that any of them could obtain by playing any of the pure strategies (which is at least 5). If communication and utility transfers among players were allowed, both players would be better-off by playing one of the pure strategy equilibria and having the player with the highest payoff and transferring 2.5 units to the other player so that both make a higher benefit of 7.5 units.

12.3 Which Equilibrium?

In the case of zero-sum games, if multiple Nash equilibria exist, they all have the same payoff. Hence, in those circumstances, the specific equilibrium the players ultimately settle for is largely irrelevant. However, as the example from the previous section shows, in non-zero-sum games, the payoffs of different equilibria can be quite different. This makes interpreting Nash equilibria and predicting the outcome of the game more difficult.

To illustrate these phenomena, consider the so-called *game of chicken*. The name has its origins in a game in which two drivers drive toward each other on a collision course: one must swerve, or both may die in the crash, but if one driver swerves and the other does not, the one who swerved will be called a *chicken*. A similar game was played by youths in the 1950s and inspired the classic James Dean movie *Rebel without a cause*.

An example of a possible payoff matrix associated with this game can be seen in Table 12.7. The payoff of -100 for each player in case of a collision is meant to represent a big loss (at least, when compared against the small profit/loss made

Table 12.7 The game of chicken.

		Hans	
		Swerve	Straight
Ileena	Swerve 1	$(0, 0)$	$(-1, 1)$
	Straight 2	$(1, -1)$	$(-100, -100)$

Table 12.8 Best responses for Ileena in the game of chicken.

If Hans...	Ileena should...
... swerves,	... go straight.
... goes straight,	... swerve (better chicken than dead!).

when one player swerves and the other goes straight). The game of chicken has been used to model a number of real-life situations, including the doctrine of mutually assured destruction during the Cold War.

Let's consider the best responses from each player. Table 12.8 shows the best responses for Ileena which, because of the symmetry of the payoff table, also apply to Hans. From this table, it is clear that there are two pure-strategy equilibria, which correspond to one of the players swerving and the other going straight. In the notation of mixed-strategy equilibria, these two equilibria correspond to $(q = 1, p = 0)$ and $(q = 0, p = 1)$, where q and p are the probabilities that Ileena and Hans swerve, respectively. Both of these equilibria imply outcomes in which there is no crash, but one of the players is always the "chicken".

Besides these two pure-strategy equilibria, the game also admits a true mixed-strategy equilibrium, which corresponds to the players swerving 99% of the time and going straight 1%. To see this, let p be the probability that Hans swerves. Table 12.9 presents the expected payoffs of the game for Ileena. Since, in the equilibrium, the utilities of swerving and going straight must be same for both options, we have

$$p - 1 = 101p - 100 \quad \Leftrightarrow \quad 99 = 100p \quad \Leftrightarrow \quad p = \frac{99}{100}$$

The expected value of the game for this mixed-strategy equilibrium is $1 - p = 1 - \frac{99}{100} = \frac{1}{100}$ for each of the players.

The outcome implied by this third mixed-strategy equilibrium is a very troubling one. Indeed, although the probability that both players go straight at the same time is really small (because of independence, 1/10,000), if the game is played for long enough the law of large numbers ensures that a crash will eventually occur!

Table 12.9 Expected utility for Ileena in the game of chicken.

If Ileena plays ...	the expected utility for Ileena is ...
... strategy 1,	... $0 \times p + (-1) \times (1 - p) = p - 1$.
... strategy 2,	... $1 \times p + (-100) \times (1 - p) = 101p - 100$.

What equilibrium will prevail in practice, assuming that the game is played multiple times and the players can learn from each other? Note that the utilities of the players are

$$U_1 = -q(1 - p) + (1 - q)p - 100(1 - q)(1 - p)$$

for Ileena and

$$U_2 = q(1 - p) - (1 - q)p - 100(1 - q)(1 - p)$$

for Hans. Let's assume that both players start the game playing according to the mixed-strategy Nash equilibrium. In particular, if Ileena sticks to the optimal strategy $q = 0.99$ and we can plot the utilities of both players as a function of p (the probability that Ileena will swerve) using the following R code:

```
> q   = 99/100
> p   = seq(0, 1, length=101)
> U1 = - q*(1-p) + (1-q)*p - 100* (1-q)*(1-p)
> U2 = q*(1-p) - (1-q)*p - 100* (1-q)*(1-p)
> plot(p, U1, xlab="p", ylab="Utilities", type="l")
> lines(p, U2, lty=2)
> abline(v = 99/100, lty=3)
```

As expected, the point where the two utilities intersect corresponds to $p = 0.99$, which is the equilibrium strategy. Furthermore, Figure 12.1 shows that, no matter what Hans does, his utility remains constant at -0.01, and there is no incentive for him not to play the Nash equilibrium $p = 0.99$. However, since the utility of Hans is constant, there is also no incentive for him to play it for sure either! Indeed, although Hans cannot increase his profit, he can reduce (or even increase!) the profit of Ileena. In the two extremes, if $p = 1$ then Hans can on its own maximize the profit of Ileena (making it 0.01), and if $p = 0$ then the profit for Ileena would be minimized (making it -1.99).

We have implicitly assumed that the outcome for Ileena does not matter to Hans: there is no reason why Hans would prefer one of these outcomes over the other. However, in real-life players might indeed have a preference for making the other player better (or worse off). For the sake of argument, let's assume that Hans decides not to play the mixed-strategy equilibrium, but instead is a bit more aggressive and slightly decrease his probability of swerving in order to reduce the payoff of Ileena. The following code can be used to plot the utility of each player as a function of q if $p = 0.9$:

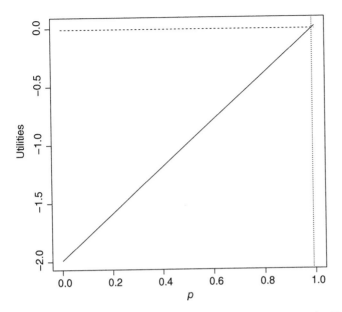

Figure 12.1 Expected utilities for Ileena (solid line) and Hans (dashed line) in the game of chicken as function of the probability that Hans will swerve with probability p if we assume that Ileena swerves with probability $q = 0.99$.

```
> q   = seq(0, 1, length=101)
> p   = 0.9
> U1 = - q*(1-p) + (1-q)*p - 100* (1-q)*(1-p)
> U2 = q*(1-p) - (1-q)*p - 100* (1-q)*(1-p)
> plot(q, U1, xlab="q", ylab="Utilities", type="l",
+       ylim=c(-10.9,0.01))
> lines(q, U2, lty=2)
```

Figure 12.2 indicates that, by adopting a slightly more aggressive strategy, Hans has completely changed the incentives for Ileena. If Ileena realizes the change in Hans strategy, then she will also surely change her strategy to $q = 1$, which is the choice that maximizes her utility. What happens then? The following code again plots the utilities of both players as a function of p when $q = 1$:

```
> q   = 1
> p   = seq(0, 1, length=101)
> U1 = - q*(1-p) + (1-q)*p - 100* (1-q)*(1-p)
> U2 = q*(1-p) - (1-q)*p - 100* (1-q)*(1-p)
> plot(p, U1, xlab="p", ylab="Utilities", type="l",
+       ylim=c(-1,1))
> lines(p, U2, lty=2)
```

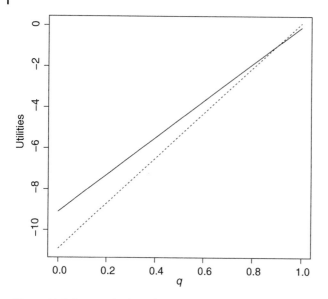

Figure 12.2 Expected utilities for Ileena (solid line) and Hans (dashed line) in the game of chicken as function of the probability that Ileena will swerve with probability q if we assume that Hans swerves with probability $p = 0.9$.

Figure 12.3 indicates that, once Ileena starts to swerve all the time, Hans will start to go straight all the time, which coincides with one of the pure-strategy Nash equilibria we identified at the beginning. But unlike the mixed-strategy equilibria, not only the players have no incentive to deviate but they also have strong incentives to stick to their strategies. A similar argument can be made if the players begin playing with the mixed-strategy equilibrium and one of the players decides to slightly increase the probability of swerving. In that case, the first player to increase the probability will see himself trapped in an equilibrium in which it will have to swerve all the time!

The previous discussion demonstrates that the mixed-strategy equilibrium of the game of chicken is *unstable*, that is, that once one of the players slightly deviates from it the steady state of the game will move toward a different equilibrium. Unstable equilibria are fragile and unlikely to persist for long in the real world. On the other hand, the pure-strategy equilibria of the game of chicken are *stable* and, once adopted, small deviations are unlikely to change the outcome of the game.

12.4 Asymmetric Games

The non-zero-sum games studied so far have been *symmetric* in the sense that both players have the same strategies and payoffs. However, not all games are

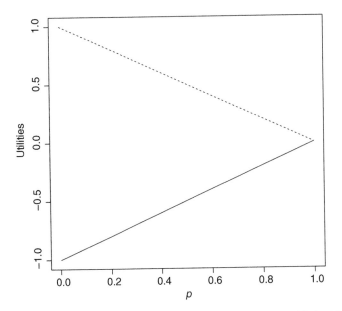

Figure 12.3 Expected utilities for Ileena (solid line) and Hans (dashed line) in the game of chicken as function of the probability that Hans will swerve with probability p if we assume that Ileena always swerves.

symmetric in their gains for both players; for example, consider a two-person fencing game where each player has only one attack move and one defensive move, but they have different gains for each player (imagine one player is a more defensive player – Ki-Adi Mundi – and the other is more skillful in offensive moves – Asajj Ventress). Table 12.10 shows the payoffs associated with this game.

Best responses for each of the players are presented in Tables 12.11 and 12.12. Again, there are two pure-strategy equilibria, one in which Ki-Adi always attacks and Asajj always defends, and another in which the roles are reversed. In addition, there is a mixed-strategy equilibrium in which Ki-Adi attacks with probability p_a and defends with probability p_d while Asajj attacks with probability q_a and defends with probability q_d.

Table 12.10 A fictional game of swords in Star Wars.

		Asajj	
		Attack move	Defensive move
Ki-Adi	Attack move	$(0,0)$	$(1,3)$
	Defensive move	$(1,2)$	$(0,0)$

Table 12.11 Best responses for Ki-Adi
in the sword game.

If Asajj...	Ki-Adi should...
... attacks,	... defend.
... defend,	... attack.

Table 12.12 Best responses for Asajj
in the sword game.

If Ki-Adi...	Asajj should...
... attacks,	... defend.
... defend,	... attack.

Table 12.13 Expected utility for Ki-Adi in the asymmetric
sword game.

If Ki-Adi...	the expected utility for Ki-Adi is ...
... attacks,	... $0 \times q_a + 1 \times q_d = q_d$
... defends,	... $1 \times q_a + 0 \times q_d = q_a$

Recall that q_a and q_d are, respectively, the probabilities that
Asajj will decide to attack or defend.

We proceed now to find p_a, p_d, q_a, and q_d. Table 12.13 presents the expected value of the game for Ki-Adi under each of this possible actions. As we have argued in the past, for Ki-Adi to be willing to randomize, the utility derived from both options has to be the same in equilibrium. This means that $q_a = q_d$ and, since $q_a + q_d = 1$, we have

$$q_a + q_a = 1 \quad \Leftrightarrow \quad q_a = \frac{1}{2}.$$

Hence, the optimal strategy for Asajj involves attacking 50% of the time, and defending the other 50% of the times.

We can carry out a similar calculation for Ki-Adi, with Table 12.14 presenting the expected value for each of his choices. Equating the expected utilities we have $3p_d = 2p_a$, or $p_d = \frac{2}{3}p_a$. Now, using the fact that $p_a + p_d = 1$ we have

$$p_a + \frac{2}{3}p_a = 1 \quad \Leftrightarrow \quad 3p_a + 2p_a = 3 \quad \Leftrightarrow \quad 5p_a = 3 \quad \Leftrightarrow \quad p_a = \frac{3}{5}.$$

Table 12.14 Expected utility for Asajj in the asymmetric sword game.

If Asajj...	the expected utility for Asajj is ...
... attacks,	... $0 \times p_a + 3 \times p_d = 3p_d$
... defends,	... $2 \times p_a + 0 \times p_d = 2p_a$

Recall that p_a and p_d are, respectively, the probabilities that Ki-Adi will decide to attack or defend.

Hence, the optimal strategy for Ki-Adi is to defend 40% of the time and to use his attacking move 60% of the time.

12.5 Exercises

1. Consider the following two-player, non-zero-sum game:

			Player 2	
		1	2	3
Player 1	A	$(2,1)$	$(1,1)$	$(1,2)$
	B	$(1,4)$	$(3,2)$	$(1,2)$
	C	$(1,2)$	$(1,2)$	$(2,1)$

(a) Does either player have a dominant strategy?
(b) Does either player have a dominated strategy?
(c) What are all the pure-strategy Nash equilibria? If there aren't any, is there a mixed-strategy equilibria?

2. Consider the following two-player, non-zero-sum game:

			Rose	
		a	b	c
Joe	X	$(4,-1)$	$(2,3)$	$(0,0)$
	Y	$(-1,2)$	$(-2,0)$	$(3,1)$
	Z	$(0,1)$	$(2,1)$	$(-2,2)$

(a) Does either player have a dominant strategy?
(b) Does either player have a dominated strategy?
(c) What are all the pure-strategy Nash equilibria? If there aren't any, is there a mixed-strategy equilibrium?

3. Consider the following two-player, non-zero-sum game:

		Romney		
		Left	Center	Right
Obama	Left	(18, 18)	(15, 20)	(9, 18)
	Center	(20, 15)	(16, 16)	(8, 12)
	Right	(18, 9)	(12, 8)	(0, 0)

(a) Does either player have a dominant strategy?
(b) Does either player have a dominated strategy?
(c) What are all the pure-strategy Nash equilibria? If there aren't any, is there a mixed-strategy equilibria?

4. Splitting three chocolate chips. Two kids are asked to write down an integer number between 1 and 3. The player with the higher number gets that amount of chocolate chips minus the amount written by the opponent (if they chose 2 and 1, respectively, the child that had chosen the higher number gets one chocolate chip). The kid who chooses the smaller number of chips will get the remainder of chips (in the example above, the second kid will get two chips). If they choose the same number of chips, they split the three chips equally (i.e., 1.5 each). Put the game in normal form. What is the solution of the game?

5. In the game described in the previous exercise, assume instead that the kid with highest number gets that amount of chocolate chips minus the amount written by the opponent, but the second opponent gets the number they asked for (to a maximum of three). First put this game in normal form. What is the solution of this game? Explain which differences you see between these two games.

6. Tax collection. Consider a game between a tax collector (Mary) and a tax payer (John). John has income of 200 and may either report his income truthfully or lie. If he reports truthfully, he pays 100 to Mary and keeps the rest. If John lies and Mary does not audit, then John keeps all his income. If John lies and Mary audits, then John gives all his income to Mary. The cost of conducting an audit is 20. Suppose that both parties move simultaneously (i.e., Mary must decide whether to audit before he

knows John's reported income). Find the mixed-strategy Nash equilibrium for this game and the equilibrium payoffs to each player.

7. In this simple coin game, players start by choosing randomly whether to have either no coin or one coin in their hand; secondly, each of them also needs to try to guess the sum of coins in their hands. They win if they get the right sum and they draw when they both have the right answer; the players loose in any other situation.
   (a) Compute the payoff table associated with this game and show that this is a non-zero-sum game.
   (b) Find all pure- and mixed-strategy equilibria for this game.
   (c) Which change to the rules of the previous game would turn it into a zero-sum game?

8. A company's cash is contained in two safes, which are kept some distance apart. There is $90,000 in one safe and $10,000 in the other. A burglar plans to break into one safe and have an accomplice set off the alarm in the other one. The watchman has time to check only one safe; if he guards the wrong one, the company loses the contents of the other safe, and if he guards the right one, the burglar leaves empty-handed. From which safe is a sophisticated burglar more likely to steal? With what probability? What should the watchman do, and how much, on average, will be stolen?

9. Why do cops separately interrogate suspected accomplices of a crime?

10. Can you give a historical example in which two (groups) of countries could be thought of as playing one of the pure-strategy equilibria in the game of chicken? How about an example of two countries playing the mixed-strategy equilibrium?

11. [R] Use the simulation of the game of chicken to investigate what happens if one player plays using the randomized strategy associated with the mixed-strategy Nash equilibrium while the other player always goes straight. Which one of the two equilibriums do you think is more likely to become the steady state after repeated play?

12. [R] Build a simulation for the asymmetric game in Section 12.4 and contrast the various equilibria.

13

Tic-Tac-Toe and Other Sequential Games of Perfect Information

For the most part, the games we have considered so far are such that both players decide their strategies simultaneously. However, games such as chess or checkers involve players who take turns. These *sequential games* are fundamentally different from the simultaneous games we studied in previous chapters because players can account for the moves previously made by their opponent when making their own decisions. For simplicity, we will focus on games of *perfect information*, in which outcomes are not randomly determined.

13.1   The Centipede Game

In the centipede game, two players (call them Carissa and Sahar) alternately face two stacks of money. In their turn, each player must choose between passing the stacks along, in which case both grow slightly and the next participant gets to play, or to take the larger stack for themselves, in which case the other player gets the smaller one and the game ends. In any case, the game ends after a predetermined number of rounds.

In this section, we focus on a three-round version of the centipede game in which Carissa gets to play first and faces two piles of coins, one containing $3 and the other containing $1. Each time a player decides to pass, the money in each pile doubles. Hence, if Carissa decides to pass on the first round, the amount of money in the stacks grows to $6 and $2, respectively, and Sahar has to choose between taking the stack of $6, or passing the turn to Carissa (in which case the amount of money on each stack will double again to $12 and $4). During the third and last round, Carissa must decide whether to take the larger stack or divide the money evenly between her and Sahar.

All possible outcomes in the centipede game can be enumerated by using a tree similar to the decision trees we discussed in Chapters 5, 6, and 11. In this tree, each level represents the different options available to the next player conditionally on the previous plays. These trees, when accompanied by the payoffs

Probability, Decisions and Games: A Gentle Introduction using R, First Edition. Abel Rodríguez and Bruno Mendes.
© 2018 John Wiley & Sons, Inc. Published 2018 by John Wiley & Sons, Inc.
Companion website: www.wiley.com/go/Rodriguez/Probability_Decisions_and_Games

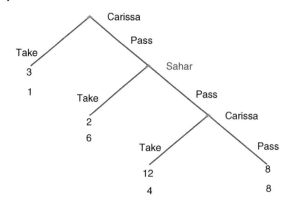

Figure 13.1 Extensive-form representation of the centipede game.

associated with each terminal node, are called the *extensive-form* representation of the game (see Figure 13.1). In the extensive-form representation of a game, each decision point corresponds to a node in the tree, while each possible decision is associated with a branch. The payoffs at the end of each branch of the tree show the utility that each player derives from the particular combination of choices that leads to it, with the first number representing the utility of the player who plays first.

Using the extensive-form representation of a sequential game simplifies its solution. In sequential games, a solution is found by determining optimal actions for each player *at each stage of the game* (including at those stages of the game that we might not rationally be expected to happen). In the centipede game we just described, this means determining what Carissa should do in the first and third rounds of the game and what Sahar should do during the second round (even though, as we will see in a minute, we rationally expect the game to end in the first round).

What should Carissa and Sahar do to optimally play the centipede game? Sahar's strategy is relatively easy to figure out: if she gets to play, she should take the largest stack. Indeed, it should be clear that if Sahar decides to pass in her turn, Carissa can be expected to grab the large stack and leave Sahar with $4 (which is worse than the $6 she could get by grabbing the largest stack in her turn). However, Carissa's strategy is less obvious: should she grab the largest stack in the first round (netting her $3) or should she wait until her next turn (in which case she will get $12)? Actually, Carissa should not be too greedy and should take the largest stack in her first round of play (which will leave her with $3 and Sahar with $1). Indeed, our previous discussion on Sahar's strategy suggests that, if Carissa passes in the first round, she will never get to play the third one! Hence, since Sahar can be expected to take the largest stack in her turn (leaving Carissa with only $2), Carissa is better off just if she just grabs the larger stack in the first round and does not let Sahar play.

The previous argument suggests that sequential games can be solved by recursively deciding what is the best decision faced by the *last* player to move and then moving up the decision tree. This procedure for solving sequential games is known as *backward induction,* and it can be used to solve any finite, two-person sequential game of perfect information.

The Backward Induction Algorithm

1. For each of the final decision nodes, solve for the optimal behavior of the players (i.e., see which is the best option for the player who plays last).
2. For each of those final decision nodes, replace the branches with the payoff associated with the best decision of that player.
3. Repeat steps 1 and 2 for this reduced game until the initial decision node is reached.

To illustrate how the backward induction algorithm works, let's apply it to the centipede game by recursively trimming the tree in Figure 13.1. Note that Carissa is the last player to act, and she faces the decision of taking the stack (which pays her $12) or passing (which pays her $8). The decision is easy: she takes the stack. Once we have figured this out, we can replace the two bottom right branches of the tree with the payoffs associated with Carissa's optimal decision on the third round, leading to the reduced tree in Figure 13.2.

The procedure can now be iterated. To solve for Sahar's optimal action during the second round of play note that, as we discussed before, she needs to decide between taking the larger stack (a profit of $6) or passing (which, as Figure 13.2 clearly shows, leads to a profit of $4). By replacing these two branches with Sahar's optimal choice (taking the larger stack), we obtain another reduced tree (see Figure 13.3). At this point, it is easy to see that Carissa should always choose to take the largest stack in the first round of the game (which gives her a payoff of $3) over passing (which would lead to a profit of $2 if Sahar plays optimally).

In summary, the solution of the centipede game is as follows:

- Carissa's optimal play is to take the larger stack in the first round of play.

Figure 13.2 Reduced extensive-form representation of the centipede game after solving for Carissa's optimal decision during the third round of play.

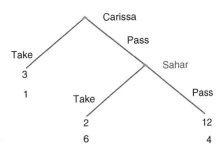

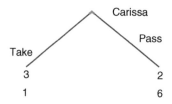

Carissa

Pass

Take

3
1

2
6

Figure 13.3 Reduced extensive-form representation of the centipede game after solving for Sahar's optimal decision during the second round of play and Carissa's optimal decision during the third round of play.

- If Carissa plays suboptimally in the first round and passes, then Sahar should take the larger stack for herself on the second round of play.
- Finally, if both players passed in the first and second rounds, then Carissa should take the larger stack on the third turn.

If both players play optimally, then Carissa will make $3 by playing this game, while Sahar will make $1.

13.2 Tic-Tac-Toe

Tic-tac-toe (also known as *Xs and Os* or *noughts and crosses*) is a two-player, pencil-and-paper game in which players take turns placing their mark (an X for one of the players and O for the other) inside the cells of a 3-by-3 grid. The winner of the game is the first player to place three of their marks in a row (either horizontally, vertically, or diagonally). If none of the players is able to place three marks in a row, the game ends in a draw. Figure 13.4 shows a sequence of plays associated with a game of tic-tac-toe in which the player represented by X plays first and the player represented by O wins the game.

Figure 13.5 shows a small part of the extensive-form representation for tic-tac-toe. Unlike the centipede game, the extensive form of tic-tac-toe is

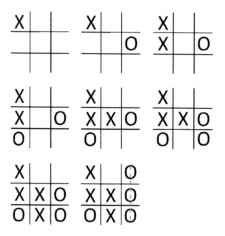

Figure 13.4 A game of tic-tac-toe where the player represented by X plays first and the player represented by O wins the game. The boards should be read left to right and then top to bottom.

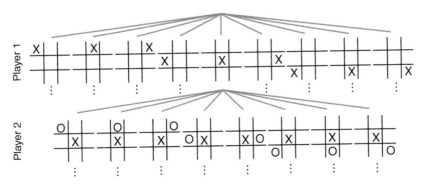

Figure 13.5 A small subsection of the extensive-form representation of tic-tac-toe.

quite unwieldy. In principle, the first player has 9 possible options for their first move, the second player can place their mark in any of 8 positions during the second round of play (since they cannot place their mark on the square already taken by the first player), and so on until one of the players wins. This suggests that the number of different games of tic-tac-toe is $9! = 362,880$. However, this number is too high: a game can finish in less than 9 moves and, once a player wins, the rest of the moves do not matter. Similarly, boards in which both players win, or where a single player wins in two different ways are of no interest.

In the end, the total number of possible final boards of tic-tac-toe is 255,168: 1,440 boards in which the X player wins after five moves have been made, 5,328 boards in which the O player wins after six moves, 47,952 in which the X player wins after seven moves, 72,576 in which the O player wins after eight moves, 81,792 in which the X wins on the ninth move, and 46,080 boards that end on a tie after nine moves. Therefore, there are $1,440 + 47,952 + 81,792 = 131,184$ possible games in which the X player wins, $5,328 + 72,576 = 77,904$ in which the O player wins, and 46,080 in which the players tie.

The computation of the number of different boards mentioned earlier uses some of the concept of combinations and permutation that we studied in Chapter 4. For example, to compute the number of games that end after the five moves, note that there are a total of eight lines of three squares (three vertical, three horizontal, and two diagonal) and it matters in which order the first player place their marks. This gives you $8 \times 3! = 48$ ways in which the three Xs can be placed. On the other hand, the two Os can be placed in any open square, of which there are six, and the order in which they are placed matters again. Hence, the total number of games that end in the fifth move is $48 \times {}_6P_2 = 48 \times 30 = 1440$. Similarly, for the number of boards that result in a draw after nine moves, note that there is a total of 16 possible patterns for the five Xs and four Os which have no three in a row. Since the order in which the Xs and Os were placed in the board matters, we are looking at $16 \times 5! \times 4! = 46{,}080$ different games. Calculations for other cases are similar but more convoluted. Alternatively, we can explicitly enumerate all possible options. The following code does that for the number of boards that correspond to a win six moves:

```
> count = 0
> for(i1 in seq(1,9)){
+    for(i2 in seq(1,9)[-i1]){
+      for(i3 in seq(1,9)[-c(i1,i2)]){
+        for(i4 in seq(1,9)[-c(i1,i2,i3)]){
+          for(i5 in seq(1,9)[-c(i1,i2,i3,i4)]){
+            for(i6 in seq(1,9)[-c(i1,i2,i3,i4,i5)]){
+              B = matrix(0, nrow=3, ncol=3)
+              B[i1] = "X"
+              B[i2] = "O"
+              B[i3] = "X"
+              B[i4] = "O"
+              B[i5] = "X"
+              B[i6] = "O"
+              Z = (B=="X")
+              Y = (B=="O")
+              if(Z[1,1]+Z[1,2]+Z[1,3]<3 & # X didn't win in 5
+                 Z[2,1]+Z[2,2]+Z[2,3]<3 &
+                 Z[3,1]+Z[3,2]+Z[3,3]<3 &
+                 Z[1,1]+Z[2,1]+Z[3,1]<3 &
+                 Z[1,2]+Z[2,2]+Z[3,2]<3 &
+                 Z[1,3]+Z[2,3]+Z[3,3]<3 &
+                 Z[1,1]+Z[2,2]+Z[3,3]<3 &
+                 Z[3,1]+Z[2,2]+Z[1,3]<3){
+                if(Y[1,1]+Y[1,2]+Y[1,3]==3 |
+                   Y[2,1]+Y[2,2]+Y[2,3]==3 |
+                   Y[3,1]+Y[3,2]+Y[3,3]==3 |
+                   Y[1,1]+Y[2,1]+Y[3,1]==3 |
+                   Y[1,2]+Y[2,2]+Y[3,2]==3 |
+                   Y[1,3]+Y[2,3]+Y[3,3]==3 |
+                   Y[1,1]+Y[2,2]+Y[3,3]==3 |
+                   Y[3,1]+Y[2,2]+Y[1,3]==3){
+                  count = count + 1
+                  #print(B) # Uncomment if you want to see
+                            # all board configurations
+                }
+              }
+            }
+          }
+        }
+      }
+    }
+  }
+ }
> count

[1] 5328
```

X		O
F	O	
X	F	X

X		F
F	X	
X	O	O

Figure 13.6 Examples of boards in which the player using the X mark created a fork for themselves, a situation that should be avoided by their opponent. In the left figure, player 1 (who is using the X) claimed the top left corner in their first move, then player 2 claimed the top right corner, player 1 responded by claiming the bottom right corner, which forces player 2 to claim the center square (in order to block a win), and player 1 claims the bottom left corner too. At this point, player 1 has created a fork since they can win by placing a mark on either of the cells marked with an F. Similarly, in the right figure player 1 claimed the top left corner, player 2 responded by claiming the bottom edge square, then player 1 took the center square, which forced player 2 to take the bottom right corner to block a win. After that, if player 1 places their mark on the bottom left corner they would have created a fork.

At first sight, the numbers presented earlier would suggest that the first player to move in tic-tac-toe has an advantage and would win about $131{,}184/255{,}168 \times 100\% \approx 51.4\%$ of the time. However, if you have ever played tic-tac-toe, you know that both players have strategies that allow them to at least draw, and maybe win, depending on how well their opponent plays the game. As with the centipede game, those strategies can in principle be found using the backward induction algorithm described in the previous section, but applying the algorithm in this case is more difficult because of the sheer size of the tree. As a consequence, we omit the details of the derivation and instead note that the description of the optimal strategy can be somewhat simplified by exploiting the fact that the board is symmetric under rotations and reflections. Indeed, the first player really has only three types of first moves: they can either place the mark in one of the corners, in one of the edges, or in the center. The second player must always respond to a corner opening with a center mark, to a center opening with a corner mark and to an edge opening with a center mark, a corner mark next to the opponent's opening mark, or with an edge mark opposite to the opponent's opening mark. Any other response from the second player would allow the first player to win. After that, players must first attempt to block any move that could lead to their opponent winning or opening a fork (a situation in which the opponent could win with two possible moves, see Figure 13.6) while attempting to create a fork for themselves or complete three marks in a row.

13.3 The Game of Nim and the First- and Second-Mover Advantages

In the game of Nim (from the German verb "nehme," which means "to take") players take turns at removing items from one or more piles, with the player removing the last piece winning the game. Nim appeared in Europe sometime during the fifteenth century. However, because of its similarities with the game of Tsyanshidzi or "picking stones" it is believed to have originated in China.

Consider a version of the game of Nim involving a stack of four pieces and two players, Ann and Mat. Starting with Ann, players take turns removing either one or two pieces from the stack; the losing player pays $1 to the winner. Figure 13.7 shows the reduced-form representation of this game. For example, note that if Ann decides to remove just one piece in the first turn, Mat decides to also remove one piece in the second round, and Ann removes one piece in the third, then Mat automatically wins (as he can remove the last piece in the stack during the fourth round). The outcome associated with other branches can be derived in a similar way.

The backward induction algorithm described in Section 13.1 can be used to derive an optimal strategy for each player and predict the outcome of the game. We start pruning the tree at the bottom left branches. At this point, Ann should choose to remove two pieces (which leads to her winning $1), which is better than removing one piece (which would lead to her losing $1). Figure 13.8 presents the corresponding pruned tree.

Now we can proceed to derive Mat's strategy during the second round of the game. If Ann removed one piece in the first round (left side of the tree), Mat is in an equally bad situation: no matter what he does, in both cases he will lose $1. On the other hand, if Ann removed two pieces in the first round, Mat should also remove two of them (as this would win him the game). The reduced tree is presented in Figure 13.9. At this point, it is clear that Ann should remove just one piece in the first round, which would win her the game.

In summary, the optimal strategy for this game is as follows:

- Ann should remove one piece in the first round. If Ann does so, Mat is indifferent between removing one or two pieces in the second round. If Mat

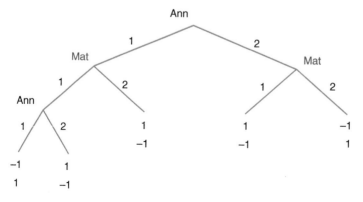

Figure 13.7 Extended-form representation of the game of Nim with four initial pieces.

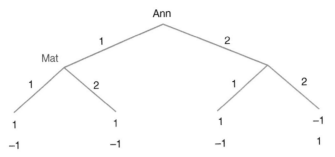

Figure 13.8 Pruned tree for a game of Nim with four initial pieces after the optimal strategy at the third round has been elucidated.

Figure 13.9 Pruned tree for a game of Nim with four initial pieces after the optimal strategy at the second round has been elucidated.

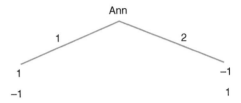

decides to remove one, Ann should remove two in the third round, winning the game. On the other hand, if Mat decides to remove two, Ann should remove one in the third round, again winning the game.

- If Ann removes two pieces in the first round, Mat should counter by removing two pieces, which would win him the game.

Note that in this version of Nim, Ann will always win as long she plays optimally. Hence, we say that Ann has a *first-mover advantage*, as she can use her first move to place herself in a position in which she cannot lose. This is because, if Ann can leave only three pieces in the table, she is guaranteed to win no matter what Mat does afterward. And since there are only four pieces to start with, removing a single piece in the first round puts Ann in that advantageous position.

It is tempting to think that the player who moves first always has an advantage. However, that is not the case. To see that this, consider now a version of Nim in which there are initially six pieces on the table instead of four. Following a similar logic than before, we can see that if Mat plays optimally, he will win no matter what Ann does. For a simple argument, note that if Mat can leave only three pieces in the table, he will win for sure. Now, if Ann starts by removing one piece in the first round, Matt can respond by removing two, which leaves three pieces in the table. On the other hand, if Ann decides to remove two pieces in the first round, Matt can respond by removing one piece, again leaving three pieces in the table. In a situation such as this, we say that Mat has a *second-mover advantage*.

13.4 Can Sequential Games be Fun?

In all the three examples we have considered so far in this chapter, it was possible to find optimal strategies for the games using backward induction and, once those strategies were obtained, the outcome of the game was predetermined. As in the case of tic-tac-toe, expert players would always draw, while in the centipede game, the first player would always make the most money and in a game of Nim with six pieces the second player will always win. This is not an exclusive feature of these games, as *Zermelo's theorem* ensures that sequential games of perfect information always have solutions that do not involve chance.

Zermelo's Theorem

In any sequential and finite two-person game of perfect information in which chance does not affect the decision making process, if the game cannot end in a draw, then one of the two players must have a winning strategy that prevails no matter what the opponent does.

Zermelo's theorem would suggest that playing sequential games of perfect information is quite boring. Indeed, once you use backward induction to *break* the game (i.e., find an optimal strategy), all you need to do is stick to that optimal strategy. If you do this, you either will always win, or will always tie, or you can only win if your opponent makes a mistake! This is the likely reason why you have not played tic-tac-toe since you were a kid!

However, note that backward induction requires that we construct the extensive-form representation of the game, and then "prune" the tree backwards until we reach the root node. Hence, backward induction is practical only if the number of options available to each player at each stage is small enough that we can actually write down all those options (as was the case with the centipede game or Nim). Even in games such as tic-tac-toe with a relatively small number of outcomes, a computer might be needed to efficiently construct an extensive-form representation of the game. When the number of possible combination of plays is very large (such as in chess), applying backward induction becomes impractical and the outcome of the game cannot be predicted with any certainty, making the game interesting.

13.5 The Diplomacy Game

As a final example, consider modeling the relationship between two countries in a game-theoretic context. For example, let's say that the relationship between the country of Zangano and the Republic of Abazi is confrontational because they are disputing an island that lies right between their territorial waters. Initially, Zangano has three different moves available: invite Abazi to negotiate a solution, ignore the issue or issue an ultimatum. On the other hand, the Republic of Abazi has its own set of choices depending of what Zangano's does in the first round. If Zangano issues an invitation for negotiation, Abazi has the choice of accepting the invitation or occupying the island. Alternatively, if Zangano decides to ignore the whole issue, Abazi will take advantage of the situation and will immediately invade the island. Finally, if Zangano issues an ultimatum, the Republic of Abazi has the option to invade the island immediately or back down and leave the island to Zangano. The extensive-form representation for this game is shown in Figure 13.10.

To solve this game, we start by looking at the bottom of the tree in Figure 13.10. Note that if Zangano offers to negotiate, the option that leads to the highest payoff for Abazi is to accept the negotiation offer. Indeed, remember that it is the second number in the pair that represents the gain for Abazi. Therefore, Abazi will be comparing a payoff of 1 (if it decides to invade) with a payoff of 2 (if they accept the negotiation offer). This observation allow us to prune the two branches on the bottom left side of the tree and substitute Abazi by the option "Negotiate" and the corresponding pair of gains (2,2) (see Figure 13.11).

Next, we move on to the next possibility (Zangano ignores the issue) and we see there is no choice for Abazi (since there is only one alternative), so we prune that last branch and substitute "Abazi" by "Invasion by Abazi" and the corresponding pair of gains (0, 1). Finally, under the ultimatum, Abazi will prefer to backdown (a payoff of -1) instead of invading (a payout of -2), therefore,

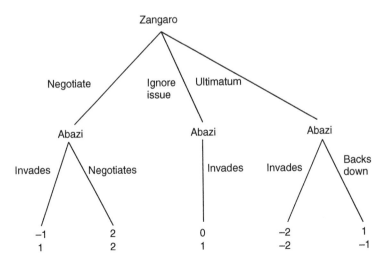

Figure 13.10 The diplomacy game in extensive form.

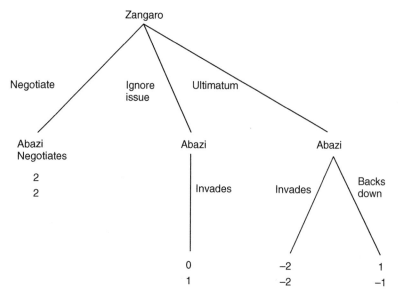

Figure 13.11 First branches pruned in the diplomacy game.

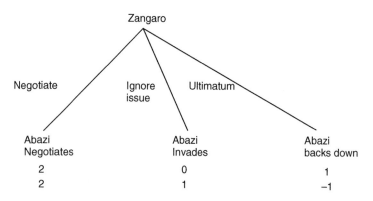

Figure 13.12 Pruned tree associated with the diplomacy game.

the last two branches are also pruned and the "Backs down" option is the only left with the corresponding pair of gains $(1, -1)$. The resulting pruned tree is shown in Figure 13.12.

To complete the solution, we need to compare the payoffs associated with Zangano's moves from Figure 13.12. Since Zangano will choose the option that maximizes its payoff (remember that Zangano's payoffs correspond to the first number on each branch), it is very clear that Zangano should choose to negotiate. Indeed, negotiating leads to a gain of 2, which is larger than the gain of 0 from ignoring the issue (which is realized when Abazi invades), or the gain of 1 from an ultimatum (since in that case Abazi backs down). In summary:

- Zangano should offer to negotiate with the Republic of Abazi.
- If Zangano does offer to negotiate, Abazi should go along and negotiate, leading to a payoff of 2 for both countries. However, if Zangano (suboptimally) decides to deliver an ultimatum, Abazi should back down, while if Zangano (again suboptimally) decides to ignore the issue, Abazi should invade.

13.6 Exercises

1. What is the backward induction method and why is it helpful?

2. What do we mean when we say that a sequential game with perfect information has been broken? Can there be any interest in playing a game that has been broken against another human? How about playing it against a computer?

3. What would the solution to the centipede game discussed in Section 13.1 be if Carissa and Sahar play the game for four rounds instead of three? What if they play it for 1,000,000 rounds?

4. What is the solution to the sequential game depicted in the figure below?

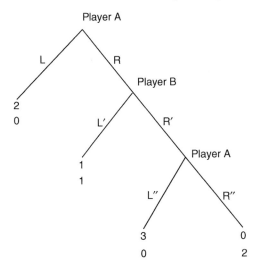

5. What is the solution for the sequential game depicted in the figure below?

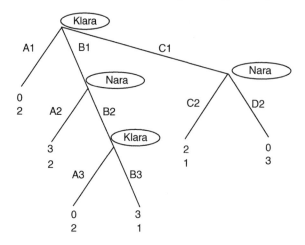

6. What is the solution for the sequential game depicted in the figure below?

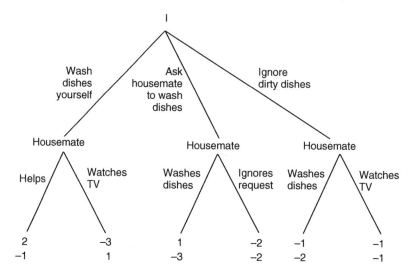

7. Construct the extensive-form representation for the game of Nim with six initial pieces that was discussed in Section 13.3, and find optimal strategies for both players.

8. A game similar to Nim. Imagine five little sticks in a row. This is a two-person game where players take turns making a choice; at each

turn, the player can remove one, two, or three matchsticks from the pile. The player who chooses the last stick loses. Can you put this game in extensive form? Find its solution.

9. What changes when you consider seven initial sticks? Discuss if it has a solution and what is it (you can put the game in extensive form if you want to).

10. The negotiation. Two persons who met through Craigslist are negotiating the price of a laptop. The seller is asking $500 for the computer (let's say this is the actual value of the computer). The buyer has the option to pay the asked price or to bargain; if the buyer decides to bargain he has the choice to *low-ball* ($250) or to ask for a 10% discount. For each of these options, the seller can either: refuse, accept or propose to split the difference ($375 for the low ball offer and $475 for the discount); this last option allows the buyer to either accept or refuse that counter-offer. If there is no agreement, both players lose $50 due to the time and effort put into meeting. Put this game in extensive form and say whether it has a solution. If there is a solution, describe it.

11. Still regarding the previous example; can you think of a simple change to the rules of the game that would make the buyer prefer to bargain?

12. Refer back to the simple two-person coin game described earlier. Let's change the game so that players take turns making a guess for the sum of the coins in their hands. If one player makes a particular choice for the sum of coins, that choice is not available to the payer who makes a decision second. Start by putting the game in extended form. What is the solution to this game?

13. Still regarding the simple coin game. Let's think about a regular gaming situation, where your opponent starts the game. Remember that you know what is in your hand but don't know what is in the opponent's hand. What is the best answer when your opponent opens the game by saying zero? (Your answer will contain different options depending on how many coins you have in your hand.) What is the best answer when the opponent says two? What is the best answer when the opponent says one?

14. Jae-Eun argues that there are $9! = 362{,}880$ possible different tic-tac-toe games, and Björn disagrees, saying that the number of different games is smaller. How could you attempt to justify Jae-Eun's statement, and why is he wrong?

15. **[R]** Write code that allows you count and print all board configurations in tic-tac-toe in which the second player wins in eight moves.

16. **[R]** Write a program that plays the game of Nim with an arbitrary number of pieces *k* optimally. Use your own code to play Nim!

A

A Brief Introduction to R

R is a freely available interactive computing environment. At its most basic, you can think of R as a fancy calculator, and you could limit yourself to using it that way. However, R offers much richer functionalities, from the ability to generate graphs to a flexible programming language that incorporates most standard mechanism for flow control. Indeed, R is a very flexible language; there are often many different ways to accomplish any given task.

Although a number of graphic user interfaces for R exist, we will be using the standard distribution, which has a command-based interface. This means that you need to communicate with the software by typing instructions into a command window. For the purpose of this book, we believe that such a stripped-down interface actually makes it easier for readers to get started. The R language is quite intuitive, so we hope that its deployment in this book will not prove an insurmountable obstacle even for readers with no programming experience.

This appendix provides readers with an introduction to the R language that covers the background needed to understand, and eventually extend, the simulations we have included in each of the book chapters. Unless you have dabbled in R before, we strongly recommend that you familiarize yourself with this appendix. If you are interested in learning more about the R environment, there are a variety of books online targeted at students with all levels of backgrounds!

A.1 Installing R

You can obtain R distribution bundles for Windows, Mac OS, and Linux at the CRAN website, https://cran.r-project.org. Just click on the link that corresponds to the appropriate operating system and follow the instructions. Once R has been installed, you can execute it by clicking on the icon. In a Microsoft

Probability, Decisions and Games: A Gentle Introduction using R, First Edition. Abel Rodríguez and Bruno Mendes.
© 2018 John Wiley & Sons, Inc. Published 2018 by John Wiley & Sons, Inc.
Companion website: www.wiley.com/go/Rodriguez/Probability_Decisions_and_Games

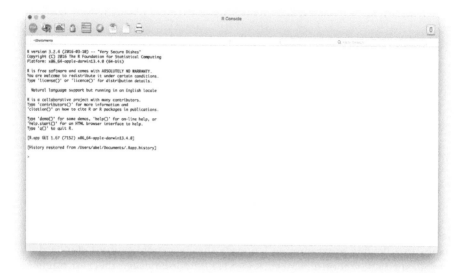

Figure A.1 The R interactive command console in a Mac OS X computer. The symbol > is a prompt for users to provide instructions; these will be executed immediately after the user presses the RETURN key.

Windows or Mac OS X computer, you should then see an interactive command console that looks like Figure A.1.

A.2 Simple Arithmetic

The last line in the command window is a prompt line that starts with the symbol >. This indicates that R is waiting for instructions. At this point, we could ask it, for example, to give us the sum of 5 and 7 by typing 5 + 7 in the prompt and then pressing the RETURN key. This is what your R console should show if you go ahead and do it:

```
> 5 + 7

[1] 12
```

Note that R provides the expected answer (the number 12) in the next line. (For now, ignore the [1] symbol at the beginning of the line, we will explain what it means later.) Similarly, you can perform many other arithmetic operations:

```
> 3*8                  # Multiplication

[1] 24

> 5^3                  # Exponentiation

[1] 125

> log10(15)            # Base 10 logarithm

[1] 1.176091

> sqrt(2)              # Square root

[1] 1.414214

> sqrt(
+ 2)

[1] 1.414214
```

The text that appears after the # symbol is a comment. We have added comments to this code to explain what the different commands do. However, you do not need to type them yourself or worry about them: any text between # and the next RETURN is ignored by R. Furthermore, as the last command illustrates, incomplete expressions (which could happen, e.g., when you press RETURN too early by mistake) are continued on the next line. Continuation lines start with the + prompt instead of the usual >.

All your standard functions (including trigonometric and exponential functions) are implemented in R. In addition to your regular arithmetic operations, you can also perform "integer" operations:

```
> 13/3                 # Regular division

[1] 4.333333

> 13%/%3               # Integer division

[1] 4

> 13%%3                # Residual of the integer division

[1] 1
```

You can get help with these functions and operators (as well as about any other R functionality) using the `help()` function. `help()` will prompt the creation of a separate window with the relevant information. For example, if you type `help("%/%")`, then a window that contains detailed help on how to carry out arithmetic operations will pop up.

The standard precedence of operations applies in R, with exponentiation being resolved before multiplications/divisions and additions/subtractions being last. However, you can use parenthesis to change the order in which operations are carried out

```
> 4 + 2*3                # Firt multiply, then sum

[1] 10

> (4 + 2)*3              # First sum, then multiply

[1] 18
```

R can treat ∞ as a number, which is represented by the symbol `Inf`. Similarly, undefined operations, such as dividing 0 by 0, return the `NaN` ("Not a Number") symbol:

```
> 3/0

[1] Inf

> 5/Inf

[1] 0

> 0/0

[1] NaN
```

A.3 Variables

You can store values in named variables that can later be used in expressions just like regular numbers. For example,

```
> x = 3
> y = 5
> z = x^2 + 2*y - x/3
```

As you could guess, variable names cannot consist of only numbers. You should also avoid the names of existing functions or operators.

To check the current value of an object, you can simply type its name at the prompt.

```
> z          # Note that 3^2 + 2*5 - 3/3 = 9 + 10 - 1 = 18

[1] 18
```

Once an object has been created, it remains in memory until you remove it (or close your current R session), so you can reuse it multiple times. You can check all objects in memory by using the command `ls` and remove an object from memory by using the function `rm`.

```
> ls()

[1] "x" "y" "z"

> rm("y")
> ls()

[1] "x" "z"
```

Some variables containing widely used constants (such as π) are already predefined:

```
> pi

[1] 3.141593

> sin(pi/6)

[1] 0.5
```

A.4 Vectors

A vector is just a list of values that share a common name but can be accessed independently of each other. You can think of a vector as a big box divided into many compartments organized sequentially, with each compartment containing a different value. You can either move the whole box around or, if needed, access the individual compartments (see Figure A.2). You can create arbitrary vectors using the c() function.

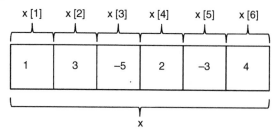

Figure A.2 A representation of a vector *x* of length 6 as a series of containers, each one of them corresponding to a different number.

```
> x = c(1,3,-5,2,-3,4)    # x is a vector with 6 elements
> x

[1]  1  3 -5  2 -3  4
```

Generally speaking, creating vectors by using the `c()` function is tedious. When the vectors follow regular patterns, you can use the `rep()` and `seq()` commands to simplify the process.

```
> x = rep(3, times=7)
> x

[1] 3 3 3 3 3 3 3

> y = seq(1, 10, by=3)
> y

[1]  1  4  7 10

> u = 1:8            # Shortcut equivalent to u = seq(1, 10, by=1)
> u

[1] 1 2 3 4 5 6 7 8

> w = rep(seq(1,5), times=8)
> w

 [1] 1 2 3 4 5 1 2 3 4 5 1 2 3 4 5 1 2 3 4 5 1 2 3 4 5 1 2 3 4
[30] 5 1 2 3 4 5 1 2 3 4 5

> z = rep(seq(1,5), each=8)
> z

 [1] 1 1 1 1 1 1 1 1 2 2 2 2 2 2 2 2 3 3 3 3 3 3 3 3 4 4 4 4 4
[30] 4 4 4 5 5 5 5 5 5 5 5
```

The meaning of the strings [1] and [36] in the first and second lines should now be clear: they tell you what is the index of the first element that appears in each line. This is meant to make it easier for you to read a vector off the screen.

You can access individual elements of a vector by using the subsetting operator [], and you can find the length of a vector by using the length() function.

```
> x = c(1,3,-5,2,-3,4)
> x[3]     # The third element in the vector x has the value -5

[1] -5

> length(x)

[1] 6
```

You can also use the [] operator to create a subvector that only contains some of the entries in the original vector. Note that negative indexes remove entries.

```
> y = x[c(2,5,6)]    # Create a subvector with three elements
>                    # corresponding to the second, fifth, and
>                    # sixth element of x
> y

[1]  3 -3  4

> w = x[-c(1,5)]     # Create a subvector with all the elements
>                    # of x except the first and fifth
> w

[1]  3 -5  2  4
```

In many ways, vectors can be manipulated as if they were scalar variables. For example, you can add or multiply two vectors of the same length. If you do, operations are carried out *elementwise*, that is, the result is another vector of the same length whose first element is the sum/product of the first elements in each of the two origianl vectors and so on:

```
> x = c( 1,3,-2, 4)
> y = c(-3,1, 5,-6)
> z = x - y
> z

[1]  4  2 -7 10
```

If the length of the two vectors is not the same, R "recycles" the entries of the shorter vector until the sizes match. This might lead to a warning or not, depending on whether the length of the longer vector is a multiple of the length of the shorter one. My recommendation is that you avoid recycling until you have gained substantial experience with R.

```
> x = c( 1,3,-2, 4)
> y = c(-3,1, 5)
> z = c(-3,1)
> w = x + y  # Operation proceeds as if y had been defined as
+               # y = c(-3,1,5,-3) (first element is recycled)

Warning in x + y: longer object length is not a multiple of
shorter object length

> x + z       # First two elements of z are recyled, no warning

[1] -2  4 -5  5
```

Many functions in R are vectorized, that is, if a vector is passed as the argument, then the function is applied individually to each element. This helps make the R code easier to read.

```
> x = c(1,3,8,4)
> log10(x)

[1] 0.0000000 0.4771213 0.9030900 0.6020600

> 2^x

[1]   2   8 256  16
```

Another example of a vectorized function is cumsum(), which provides cumulative sums of the elements of a vector. This is particularly useful if the entries of the original vector represent the payoffs of a repeated bet, in which case the cumulative sum represents the running profit/loss that the player has incurred.

```
> x = c(1,3,8,-4)
> cumsum(x) # First entry is 1, second is 1+3, third is 1+3+8,

[1]  1  4 12  8
```

Some functions are not vectorized, but are instead designed to operate on all elements of the vector simultaneously. For example, the functions sum(), `mean()`, `max()`, and `min()` give you the sum, the average, the maximum, and the minimum of all the entries of a vector,

```
> sum(x)        # Returns sum of the values in x

[1] 8

> mean(x)       # Returns the simple average of values in x

[1] 2

> min(x)        # Returns the smallest value in x

[1] -4

> max(x)        # Returns the largest value in x

[1] 8
```

A.5 Matrices

Matrices are similar to vectors, but instead of storing elements sequentially they do so in a rectangular array. Hence, entries on a matrix are indexed by two numbers; the first one corresponding to the row on which it is located; and the second corresponding to the column. Furthermore, each row or column of a matrix is simply a vector.

You can create a matrix by starting with a long vector and then using its elements to fill the matrix sequentially by either row or column.

```
> A = matrix(c(1,2,3,4,5,6), nrow=3, ncol=2)
> A            # Filled by column (the default)

     [,1] [,2]
[1,]   1    4
[2,]   2    5
[3,]   3    6

> A = matrix(c(1,2,3,4,5,6), nrow=3, ncol=2, byrow=T)
> A            # Filled by row

     [,1] [,2]
[1,]   1    2
[2,]   3    4
[3,]   5    6
```

Note that the strings [1,], [2,], and [3,] at the beginning of each line serve to identify the rows of the matrix, while the strings [,1] and [,2] identify the columns. As this suggests, the elements of the matrix can be accessed using the [] operator with two indexes separated by a comma. If you want to access a whole row or a whole column of the matrix, leave the index empty (the result will be treated as a vector).

```
> A = matrix(c(1,2,3,4,5,6), nrow=3, ncol=2)
> A

     [,1] [,2]
[1,]   1    4
[2,]   2    5
[3,]   3    6

> A[3,2]

[1] 6

> A[3,]

[1] 3 6

> A[,2]

[1] 4 5 6
```

Sometimes, it is useful to compute rowwise or columnwise sums of the elements of a matrix. The functions rowSums() and colSums() allow you do to exactly that

```
> rowSums(A)    # Rowwise sum

[1] 5 7 9

> colSums(A)    # Columnwise sum

[1]  6 15
```

More general functions can be used on each row or column of the array through the apply() function.

```
> apply(A,2,sum)        # Another way to do columnwise sums

[1]  6 15

> apply(A,1,min)        # Rowwise minimum

[1] 1 2 3

> apply(A,2,cumsum)     # Columnwise cumulative sum

      [,1] [,2]
[1,]     1    4
[2,]     3    9
[3,]     6   15
```

A.6   Logical Objects and Operations

So far, we have only discussed variables that contain real numbers. However, R allows for variables that contain other types of objects. One example corresponds to logical variables, which take only two values (TRUE and FALSE) and are the centerpiece of Boolean algebra.

Logical values are often the result of comparisons between other types of objects:

```
> 4 <= 2                # Is 4 less or equal than 2?

[1] FALSE

> 5/2 == 9/3 - 0.5      # Is 5/2 the same as 9/3 - 0.5?

[1] TRUE
```

Note that, while = is the assignment operator used to assign values to variables, == is the *equal to* operator involved in comparisons

```
> 4 == 2        # OK:  Is 4 equal to 2?

[1] FALSE

> 4 = 2         # NO:  Assign the value 2 to 4. Leads to error.

Error in 4 = 2: invalid (do_set) left-hand side to assignment
```

You can combine results from various comparisons using the and and or operators, which in Boolean algebra play a similar role to products and additions in standard algebra:

```
> 4 < 2 & 9 > 3        # ''and'' operator; both must be TRUE

[1] FALSE

> 4 < 2 | 9 > 3        # ''or'' operator; only one needs to be TRUE

[1] TRUE

> !(4 < 2)             # ''not'' operator; flips the result

[1] TRUE
```

Just like multiplications are resolved before sums by convention, and operations are resolved before or operations. As before, you can use parentheses to change the order in which operations are carried out:

```
> 4 < 2 & 4==3 | 9 > 3  # First resolve ''and,'' then the ''or''

[1] TRUE

> 4 < 2 & (4==3 | 9 > 3) # First resolve ''or,'' then the ''and''

[1] FALSE
```

Comparison operations are also vectorized:

```
> x = c(-1,2,3,1,4,6,-8,2)
> y = (x >= 2.5 & x < 6)
> y

[1] FALSE FALSE  TRUE FALSE  TRUE FALSE FALSE FALSE
```

We can check whether a variable takes at least one value among a list of possibilities by combining multiple comparisons using or operators:

```
> x = c(-1,2,3,1,4,6,-8,2)
> y = (x==1 | x==5 | x==7) #Which numbers are either 1, 5 or 7
> y

[1] FALSE FALSE FALSE  TRUE FALSE FALSE FALSE FALSE
```

However, this approach can be impractical if the number of options is large. As an alternative, we can use the `%in%` function.

```
> x = c(-1,2,3,1,4,6,-8,2)
> y = x %in% c(1,5,7)
> y

[1] FALSE FALSE FALSE  TRUE FALSE FALSE FALSE FALSE
```

The functions `any()` and `all()` provide convenient ways to check if at least one or if all the elements of the vector are true.

```
> any(y)       # Is at least one element TRUE?

[1] TRUE

> all(y)       # Are all elements TRUE?

[1] FALSE

> all(!y)      # Are all elements FALSE?

[1] FALSE
```

When arithmetic functions are used with logical vectors, TRUE values are treated as 1s and FALSE are treated as 0s.

```
> sum(y)              # Number of TRUE values in y

[1] 3

> sum(y)==length(y)   # Same as all(y)

[1] FALSE

> sum(y)>0            # Same as any(y)

[1] TRUE
```

Logical vectors provide another way to select entries of a vector. For example, if we are interested in the sub-vector of x that contains the entries that are greater than 2.5:

```
> x = c(-1,2,3,1,4,6,-8,2)
> x[x >= 2.5]

[1] 3 4 6
```

A.7 Character Objects

Characters in R are distinguished by the fact that they are enclosed in quotation marks (either single or double quote delimiters can be used, but double quote are generally preferred). You can create character vectors and perform comparisons with them just like you did with numeric vectors.

```
> x = c("Heads", "Tails")
> x

[1] "Heads" "Tails"

> z = c("CA", "NE", "OR", "OR", "CA", "UT", "CA", "OR", "NE")
> z == "CA"

[1]  TRUE FALSE FALSE FALSE  TRUE FALSE  TRUE FALSE FALSE

> z > "NE"     # Comparison based on alphabetical order

[1] FALSE FALSE  TRUE  TRUE FALSE  TRUE FALSE  TRUE FALSE
```

Arithmetic operations are not defined for character vectors, even if they only contain numbers:

```
> x = c("1", "-2", "3", "4")
> y = c("-4", "7", "1", "-2")
> x + y

Error in x + y: non-numeric argument to binary operator
```

However, you can coerce characters that only contain number to numerical objects for which regular algebraic operations are defined using the as.numeric() function.

```
> as.numeric(x) + as.numeric(y)

[1] -3  5  4  2
```

A.8   Plots

You can use R to easily create plots. For example, suppose that we want to plot the parabola $f(x) = x^2 - 2x + 1$ in the interval $[-1, 2]$. To do so, we need to first compute the value of $f(x)$ over a fine grid of values in the interval of interest. The function plot() then can be used to generate a new window that contains a Cartesian coordinate system and a series of dots that represent the coordinates of each point in the grid and the corresponding value of $f(x)$ (see Figure A.3).

```
> x = seq(-1,2,length=200)   # A regular grid with 200 values
> y = x^2 - 2*x + 1          # Function evaluated at the grid
> plot(x, y)                 # Generates the plot
```

Figure A.3 uses dots to represent the function. However, in this case, it would be more convenient to connect the values using lines. This can be easily achieved using the type option. Similarly, you can change the labels of the axes using the xlab (for the x-axis label) and the ylab (for the y-axis label) options (see Figure A.4).

```
> plot(x, y, type="l", xlab="x axis", ylab="y axis")
```

The plot function admits a number of additional parameters that are helpful in fine tuning graphs. Examples include col (which allows you to change the color of the lines/points) and lty (which allows you to use dashed and dotted lines). A full discussion of all options, however, is beyond the scope of this introduction.

When creating plots, it is usually a good idea to add reference lines that help focus attention on the features of the graph that are most relevant for the discussion at hand or to place mutliple plots on a single graph. The function abline() allows you to add *straight* reference lines to an existing plot that was previously created using the plot() function. Similarly, the functions lines and points can be used to add additional plots to an existing one. Figure A.5 was created using the following code.

```
> w = x^3 - x + 1
> plot(x, y, type="l", xlab="x axis", ylab="y axis")
> lines(x, w, lty=2)      # Second dashed curve
> abline(h=1.2, lty=3)    # Horizontal dotted line at 1.2
```

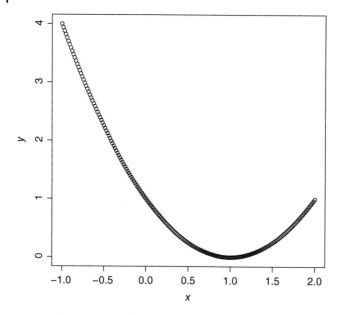

Figure A.3 An example of a scatterplot in R]An example of a scatterplot in R.

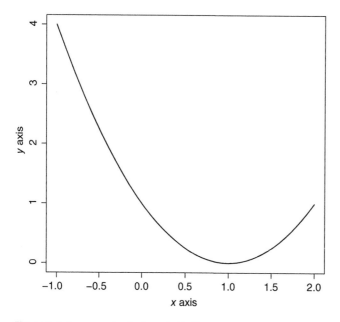

Figure A.4 An example of a line plot in R.

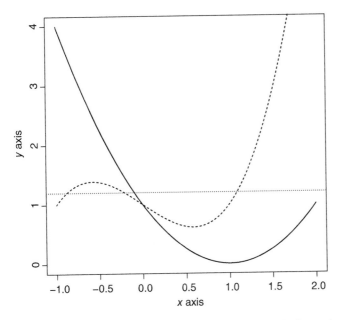

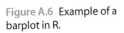

Figure A.5 Adding multiple plots and reference lines to a single graph.

Figure A.6 Example of a
barplot in R.

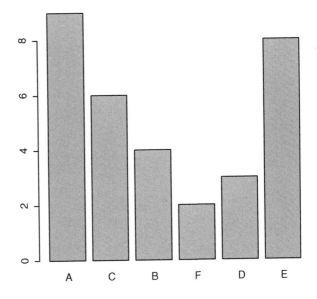

One last type of graph that will be useful as you move along the book is a bar graph. As the name suggests, in a bar graph, a list of numerical values of variables are represented by the height of rectangles of equal width. The function `barplot()` can be used to create a bar chart in R (see Figure A.6):

```
> x = c(9,6,4,2,3,8)
> coln = c("A","C","B","F","D","E")
> barplot(x, names.arg=coln)
```

A.9 Iterators

When the same operation needs to be repeated a large enough number of times, sequentially inputing the commands by hand is impractical. Vectorization sometimes offers a way to deal with these situations, but it is not always possible or practical. For example, when the outcome of one iteration depends on the results from previous ones, vectorization is usually not helpful. Loops provide a flexible alternative to deal with iterated operations.

To motivate loops, consider creating a matrix with 10 rows, each corresponding to sequences of 6 integers, all with the same starting value but different increments (increments of 4 for the first row, increments of 5 for the second, etc.). This can be achieved using the following code:

```
> A = matrix(0, nrow=10, ncol=6)
> A[1,]  = seq(1, by=4,  length=6)
> A[2,]  = seq(1, by=5,  length=6)
> A[3,]  = seq(1, by=12, length=6)
> A[4,]  = seq(1, by=3,  length=6)
> A[5,]  = seq(1, by=9,  length=6)
> A[6,]  = seq(1, by=1,  length=6)
> A[7,]  = seq(1, by=2,  length=6)
> A[8,]  = seq(1, by=6,  length=6)
> A[9,]  = seq(1, by=10, length=6)
> A[10,] = seq(1, by=8,  length=6)
> A
```

	[,1]	[,2]	[,3]	[,4]	[,5]	[,6]
[1,]	1	5	9	13	17	21
[2,]	1	6	11	16	21	26
[3,]	1	13	25	37	49	61
[4,]	1	4	7	10	13	16
[5,]	1	10	19	28	37	46
[6,]	1	2	3	4	5	6
[7,]	1	3	5	7	9	11
[8,]	1	7	13	19	25	31
[9,]	1	11	21	31	41	51
[10,]	1	9	17	25	33	41

Note that the 2nd to the 11th instructions are structurally identical. They only differ on two features: the index of the row increases and the by argument changes to reflect the desired increment in the sequence. for loops allow you to accomplish the same task without having to write one separate instruction for each row of the matrix. for loops, which allow you to repeat the same set of instructions a fixed number times, have the following syntax:

```
for(counter in vector){
    block of instructions to be repeated
}
```

The counter, which is defined within the parentheses that follow the for instruction, is a variable that sequentially takes the values contained in vector. Roughly speaking, this is the variable that tells you how many times the operations are going to be repeated. On the other hand, a set of instructions that are going to be repeated, once for every value in vector, are located within the curly brackets that follow the parentheses.

As an example, the following code uses a for loop to complete the task of filling out the rows of a matrix with different sequences of numbers:

```
> A = matrix(0, nrow=10, ncol=6)
> increments = c(4,5,12,3,9,1,2,6,10,8)
> for(i in 1:10){
+    A[i,] = seq(1, by=increments[i], length=6)
+ }
> A
```

	[,1]	[,2]	[,3]	[,4]	[,5]	[,6]
[1,]	1	5	9	13	17	21
[2,]	1	6	11	16	21	26
[3,]	1	13	25	37	49	61
[4,]	1	4	7	10	13	16
[5,]	1	10	19	28	37	46
[6,]	1	2	3	4	5	6
[7,]	1	3	5	7	9	11
[8,]	1	7	13	19	25	31
[9,]	1	11	21	31	41	51
[10,]	1	9	17	25	33	41

Iterations of a loop can depend on the result of previous iterations. For example, consider computing the first 20 terms of the *Fibonacci* sequence[1]:

---

1 The Fibonacci sequence has featured prominently in a number of movies as TV shows (including the Da Vinci code). Each term is constructed by adding together the previous two terms. The two initial terms of the recursion are both equal to 1.

```
> n = 20
> fibon = c(1,1,rep(0,n-2))
> for(i in 1:(n-2)){
+     fibon[i+2] = fibon[i+1] + fibon[i]
+ }
> fibon
```

```
 [1]    1    1    2    3    5    8   13   21   34   55   89
[12]  144  233  377  610  987 1597 2584 4181 6765
```

while loops are an alternative to for loops. Rather than being executed a fixed number of times, while loops are executed indefinitely until a given condition is satisfied. The syntax for a while loop is

```
while(condition){
    block of instructions to be repeated
}
```

The expression that replaces the placeholder condition must result in a single logical value (while loops *are not* vectorized). As before, the block of instructions that will be repeated until the condition is satisfied is placed between curly brackets. The condition associated with a while loop is checked before each iteration is executed. Hence, if the condition is not satisfied before the loop starts, the instructions inside are never executed.

As an example of the use of while loops, consider the problem of generating the first term of the Fibonacci sequence that is greater than 1000 (recall from our previous example that the value of such a term is 1597). Since we do not necessarily know in advance how many terms will need to be computed, we use a while loop that checks on the value of the Fibonacci sequence after each iteration and terminates if the current term is greater than 1000.

```
> termminus2 = 1
> termminus1 = 1
> term = termminus1 + termminus2
> while(term <= 1000){
+     termminus2 = termminus1
+     termminus1 = term
+     term = termminus1 + termminus2
+ }
> term
```

```
[1] 1597
```

A.10   Selection and Forking

You might sometimes find that different pieces of your code need to be executed depending on whether specific conditions are satisfied. For example, you might want to set the value of a variable differently depending on whether another variable is positive or negative. if/else statements allow you to accomplish this goal. The syntax for an if/else loop is

```
if(condition){
   block of instructions if condition is TRUE
}else{
   block of instructions if condition is FALSE
}
```

As with a while loop, the expression that replaces the placeholder condition must result in a single logical value. Depending on whether condition is TRUE or FALSE, only the top (or bottom) block of instructions will be executed. If an else statement is not included, then no instructions are executed when condition is FALSE.

The following code shows an example of conditional execution:

```
> x = 3
> if(x>0){
+    y = 2*x
+ }else{
+    y = x - 4
+ }
> y       # x is positive, so only the first branch is executed

[1] 6
```

if/else statements can be particularly useful in conjunction with for and while loops. The function ifelse() is a vectorized version of the if/else, but we will rarely use it in this book.

A.11   Other Things to Keep in Mind

Once you have finished with your work, you can save all of it by using the option Save Workspace File... in the Workspace menu. This will prompt a window where you can type a name for the workspace and select a folder where it will be stored. To load the workspace at a later time, you can either double click on the workspace file or use the option Load Workspace File... in the same Workspace menu.

One of the key features of R is its extendibility. A number of authors have developed groups of specialized functions that are distributed in the form of "packages". A large number of packages are available from the CRAN website. In this book, we employ the "prob" package developed by G. Jay Kern at Youngstown State University. To install the package, you can use the Package Installer option of the Packages & Data menu. Alternatively, you can use the install.packages() function from the command line.

```
> install.packages("prob")
```

In either case, you will see a number of messages associated with the installation appear in the command windows. In most circumstances, you can ignore these messages. Once the package has been installed, you will need to load it at the beginning of every R session by using the library() function:

```
> library("prob")
```

Failing to load the package before using any of its functions is a common source of errors and confusion. Please do not forget to do so!

Index

Probability, Decisions and Games: A Gentle Introduction using R, First Edition. Abel Rodríguez and Bruno Mendes.
© 2018 John Wiley & Sons, Inc. Published 2018 by John Wiley & Sons, Inc.
Companion website: www.wiley.com/go/Rodriguez/Probability_Decisions_and_Games